THE CONDITION OF RURAL SUSTAINABILITY

To Jack Marsden, my grandfather, for teaching me how to dig.

European Perspectives on Rural Development

European Perspectives on Rural Development is a series launched by the Circle for Rural European Studies (CERES). The heterogeneity and specificity of Europe's rural settings and the consequent need for well-elaborated rural development strategies provide the general framework for the series. Within this framework, local initiatives, the challenge of diversity, the need for sustainable development, the problems and perspectives of marginal areas and new responses to the current crisis in agriculture are some of the highly significant issues that will be dealt with in forthcoming volumes. The series aims to contribute to the theoretical debate on rural development within Europe and to solving the wide range of practical issues and problems now confronting many rural regions.

Several of the leading European agricultural universities, faculties and research institutes participate in the CERES network. Amongst these are Wageningen Agricultural University, the Netherlands; the University of Trás-os-Montes and Alto Douro in Vila Real, Portugal; the Escuela Tecnica de Ingenieros Agronomos of the University of Cordoba, Spain; the Universities of Cardiff, Newcastle-upon-Tyne and Hull, Great Britain; the Universities of Galway and Dublin, the Republic of Ireland; the Mediterranean Agronomic Institute in Chania, Crete, Greece; the Universities of Perugia and Pisa and the CESAR Institute in Assisi, Italy; the University of Gödöllö, Hungary; and the Johann Wolfgang Goethe University, Institut für Ländliche Strukturforschung (ILS), Frankfurt, Germany.

The Condition of Rural Sustainability

Terry Marsden

2003 ROYAL VAN GORCUM

NUR 940

ISBN 90 232 3881 8

DTP: Ans van der Lande, Wageningen
Print: Royal Van Gorcum, Assen, The Netherlands

Table of Contents

Acknowledgements

This book is very much a synthesis and reflection of empirical and theoretical work that I have been engaged in over recent years. This has involved very fruitful (and enjoyable) collaboration with a host of colleagues and friends.

In Chapters 2, 3 and 4, I am indebted to Ian Drummond and Salete Cavalcanti, Susan Andreatta and David Watts, Alberto Arce and Norman Long, for their support and collaboration. Chapters 5, 6, 7 draw upon work that I conducted with my longstanding colleagues in the UK over the past decade, namely Philip Lowe, Richard Munton, Andrew Flynn, and especially Jon Murdoch. Chapters 8–12 have been assisted by discussions and projects involving especially Jo Banks, Eduardo Sevilla Guzman, Henk Renting, Everard Smith and Mara Miele. A special thanks goes to Jan Douwe van der Ploeg, both for his patience as an editor to this series and as a supportive and creative research colleague.

I am also indebted to Joek Roex and to Ans van der Lande-Heij for their dedication to the editing and organisation of this text.

The variety of empirical work on which this book draws was supported by a range of research organisations. Of particular significance has been a series of projects funded by the Economic and Social Research Council in the UK, the European Commission FAIR project The socio-economic impact of rural development policies: realities and potentials'; and Conselho Nacional de Desenvolvimento Científico e Tecnológico in Brazil.

As always, I wish to thank Mary Anne, Joseph and Hannah for their support and patience when I was either 'miles away' conducting research, or 'miles away' at home thinking through the ideas and results of it.

Preface

Agricultural and rural development in flux: exploring the barriers and opportunities in rural sustainability

Mainstreaming the marginal

To say that the nature of agriculture, and its role in rural development is at something of a crossroads is both to understate and to reaffirm many of the debates that have been articulated in both academic and policy-making circles for more than a decade. This book analyses as much about why it is such a protracted and contested process, seemingly unresolved for so long. I take the starting point of defining the current period as a period of the 'condition of rural sustainability'; a period after the last positive vestiges of the post-war compromise associated with Fordist agriculture have been severely diminished; but a period of genuine uncertainty about what may constitute a similarly coherent and strategic set of policy and public goals for agriculture and its food supply system. The book also holds a significant proposition: *can an alternative, more sustainable rural development paradigm gain a new predominance and continuity?* If so, how will it achieve this? And what sorts of social science knowledge, expertise, politics and power frameworks will be needed in order for it to be sustained? I argue that we have little alternative; but that the road towards a more sustainable rural development model is and will not be an easy one. The book explores three social science perspectives concerning the obstacles and opportunities inherent in progressing the condition of rural sustainability. Of particular attention in this volume is the construction and exploration of a set of social scientific principles which will be needed in order to really progress towards sustainable rural development.

Following on from the more general text on the 'Condition of Sustainability' (which I co-authored with Ian Drummond in 1999), this volume takes a particular rural stance to the question of the unsustainability of present conditions. In the first two sections of the book I use particular cases to analyse the social and regulatory conditions which currently predominate in the agro-food sector. This is followed by the realisation of social alternatives, and the need for rural social scientists to develop more robust models of not only 'what is' but also 'what might be'. As in the development of the unsustainable (Section 1) it is imperative that rural social scientists begin to play more of a key role in the development of viable alternatives, through the critical analysis of both the obstacles and opportunities. This means, more substantially outlining better and more robust models of 'what may work'.

This is the spirit and purpose of this book. It is a critical treatise which struggles to find ways of making rural spaces more environmentally, socially and economically sustainable in a re-globalising world which often tends to further marginalise, devalue, bypass and exploit agrarian space and nature.

How social scientists, policy-makers, civil society can assemble the means to seriously challenge these tendencies is the key question addressed. As we shall see, this takes us beyond the modernist and post-modernist critiques of agrarian political economy, which have been so dominant in the past twenty years, to explore some new conceptual and theoretical parameters which help to build rural sustainability, almost from first principles. In particular, we need, it will be argued, a much more refined and detailed understanding of the *social regulation and governance of* rural space, in ways which not only critically examine existing conditions, but also show meaningful ways in which things *could begin to change*. Thus far, it is argued here that current and past research has not sufficiently been sensitive to the analysis of social change in its formulations. In order to incorporate this, it is necessary to completely break with the (somewhat spurious) coherence of former approaches in political economy and re-build in ways which allow change and social progressiveness a chance.

These theoretical issues and deliberations will be developed in depth at both the beginning and the end of the volume. The main sections of the book follow the three principle organising and analytical frameworks for agro-food and rural development. These are: the agro-industrial model; the post-productivist model and the more emergent sustainable rural development model. All three are co-evolving and competing at the same time; and it is important, as this analysis outlines, to delineate each in considerable detail, if the condition of rural sustainability is to be progressed. These social science perspectives are in their different ways influencing both the barriers to and opportunities for rural sustainability. Moreover, rural social scientists, by adopting these different approaches, can variably progress the condition of rural sustainability. The three models are dynamic, not only as social-science perspectives, but as analytical organising frameworks which can justify particular types of political and economic action. Rural social science is thus unable to detach itself from these dynamic tendencies.

The three subsections also make empirical reference to various types of rural space in which the author has researched over the past decade. These include Brazil and the Caribbean in 'the south', and England and

Wales in the UK. Reference is also made to other European rural regions where work is currently ongoing.

Chapter 1 starts by outlining the framework for the rest of the book by exploring the current debates surrounding food, nature and rural development from a revised regulationist perspective.

We then look at the agro-industrial, post-productivist and rural development models and perspectives in turn, over the course of the next nine chapters. The concluding chapter looks at how shifts may now occur in the move towards sustainable rural development; and the frameworks needed concerning the building of a revised developmental or ecologically modernising paradigm and the role of rural social scientists in this.

1 Introduction. Food, nature and the rural economy: regulating rural uneven development

Introduction

The governance and the regulation of nature in the rural context is reaching a point not just of crisis but also of entrenched divergence of purpose. I outline here some of the current reasons for this. They represent a slowly developing crisis of policy-making, and the expertise (not least social science expertise) relevant to that policy-making. In studying these issues we have to consider the (often contradictory) ways, in which rural nature is 'socially managed.' That is, how is it that the actions of the state, market and civil society are grouped together in seemingly legitimate ways to create particular exploitation patterns of rural nature? This social management lies at the heart of sustainable development debates. And an exploration of it gives more credence to the degree to which we can understand both the unsustainability of present conditions and the potential sustainability of future scenarios.

The foregoing analysis by way of introduction outlines three contested, but highly pertinent models of the social management of rural nature (the agro-industrial, post-productivist and rural development models) currently being applied in the advanced world. The degree to which rural sustainable development will occur will depend upon the emergence of the third development model, which at the moment still remains marginal, at least in the European sense. The durability of the rural development model will depend upon the construction of effective and legitimate governance structures and processes being established (particularly at the regional level) and the articulation of critical social science analyses which demonstrate the value of such a model for sustaining rural resources rather than upholding their unsustainability.

To say that the governance of rural space in the European context is at something of a crossroads is, of course, to understate a set of conditions that have been evolving for at least the past decade. Despite numerous policy documents and initiatives (like the Agenda 2000, rural regulation statements and the recommendations mid-term review of the CAP (2002)), there continues to be considerable difficulties concerning the development of a truly reformed European approach. For instance, despite over fifteen years of debate and policy crisis concerning the negative effects of productionist support mechanisms within the CAP, and the need to shift the emphasis towards new social and environmental agenda (including the new rural development regulation), it is still the case that the main pillar of the CAP remains in this area in terms of funding. Moreover, it still tends to reinforce the logic of agricultural productivist scale economies by rewarding the largest volume producers, as well as 'locking-in' many of the less productive producers and those least able to meet the demands such policy-designed 'technological treadmills' require.

The most contemporary manifestation of this highly asymmetrical support system is associated with the outcomes of the AGENDA 2000 CAP reform package and the establishment of what has become to be regarded as the 'second-pillar' of rural development. Whilst again this recognises the centrality of a broader sustainable rural development approach to the European countryside and its policies, the amounts of funding in the 'Second Pillar' currently remain marginal. For instance, while in 1997/8 England received over 3,625 million pounds in commodity support, representing 25,000 pounds per holding; the forecasts for 'second-pillar' funding in its first year (2001-2) is estimated as only in the order of 22 million pounds, rising to 70 million by 2006/7. A recommended target of 10 per cent of overall CAP funds is now being dismissed from 2004 onwards. Whilst this does not take into account all the impacts of much of the recent regionally allocated Structures funding, it nevertheless suggests that however urgent the calls are for a serious reform of public funds away from the productivist logic, and towards a more 'multi-functional' or orchestral approach to rural policy, the inertia in this tendency will require alternative strategies associated with different types of state, community and market-based action. Such inert tendencies at the macro-policy making level, and at some Member-state levels, represents one feature, I will argue, of a growing contradiction which lies at the heart of current rural development policy discourse. It is a rural development discourse, which to put in mildly, faces a long-term crisis of legitimacy; not necessarily amongst the rural populations of Europe alone, but also amongst the urban consumption classes.

In terms of the growing food consumption concerns, health, welfare, employment issues, this current system of agricultural and rural regulation in Europe is unsustainable. The question which I will address here is how is it that these unsustainable conditions can be, in themselves, sustained for so long? And what sort of alternatives are there that at the moment seem to remain marginal but may provide new pathways out of the impasse?

In examining the sustainability of unsustainable conditions it is necessary, I will argue, to explore three social/political/scientific/economic dynamics which are currently combining and competing in shaping European rural space. These are, *the agro-industrial model; the post-productionist model* and as yet an emerging and *alternative/sustainable rural development model.* These models of development are all variably insinuated into the current policy and scholarly debates concerning rural development, and particularly the renegotiated role that agriculture could or should play in it. They are, in this sense each struggling for primacy in the policy development process. In conclusion, I make some suggestions how these indeed may play themselves out with regard to rural Europe. It is critical to recognise how these dynamics do create spatial and social impacts which in turn condition the degree we can move from the foreclosing corners of the unsustainable to the vistas of the sustainable.

This chapter is divided into three sections. First I outline some to the principal rural governance questions which currently confront rural Europe before then detailing the nature of the three dynamics themselves. In conclusion, I will outline some of the spatial and policy implications of the third model : that of sustainable rural development.

The governance of food and rural systems in a period of crisis of legitimacy

The major and continuing pervasive influence upon rural development and particularly the role of agriculture and food within it lies fairly squarely with what can be termed the *agro-industrial dynamic.* As Table 1.1 suggests, this is associated with the globalised production of standardised products and the recent attempts to 'deregulate' international markets so as to supposedly produce more level-playing fields on which agri-business and intensive producers can compete. This is a dynamic which ties agro-food centrally into an industrial dynamic, treats natural food products as industrial products, and tends to see the reform of CAP as a preface for further concentration and global competition in the agro-food sector. This is by far the major development paradigm governing agro-food production in Europe at the present time despite nearly twenty years of attempted reform on the basis of reducing its environmental and social effects. Most of the major commodity sectors, particularly in North-western Europe and in parts of Spain and Italy are governed by what I call this 'race to the bottom' paradigm.

The case of the dairy sector is instructive here. When the Agenda 2000 negotiations started in 1998, the British government gained some supporters (Italy, Denmark, and Sweden to form the so-called 'London club') for the policy of abolishing milk quotas (by 2006). Although this has been delayed in the more recent (Berlin) summits, price cuts (15 per cent over three years) and relaxation of quota is the general tendency. This is serving to heighten the need for farm concentration and reduce still further mixed farming methods involving dairying.

Table 1.1 The three competing rural development dynamics

The Agro-industrial dynamic	The post-productivist dynamic	The rural development dynamic
Standardised products	Rural space as consumption space	Integration
Capital intensity	The marginalisation of agriculture : declining industry	Re-embedded food supply chains
Optimum (quantitative level) of production	Agriculture's share of national income falls from 2.9% in 1970 to 1.0% in 1998	New policy support structures
Long/complex supply chains	Rural land as a development space	Associational designs and networks
High levels of public funding	Social exclusion	Revised combinations of nature/value/region and quality
Continual development of 'technological fixes'	Public sector services	Rural development as counter movements
Decreasing value of primary produce and production structures	The social economy and the use of the natural as an attractor in the counter-urbanisation process	Rural livelihoods/fields of activity/new institutional arrangements
Economies of scale		Agro-ecological research and development
Rural space as agricultural space		Co-evolving supply chains
Private-interest regulation (led by retailers)/public interest regulation: crisis management and nature management		Revised state/market/civil society/nature relations
		Evaluation paradigm for rural sustainability

Globally, this industrialised and globalised competitive model of development is about the survival of the fittest and the further application of intensive technologies which reduce input costs through scaling up. However, as a telling scenario (outlined by Agra-Europe) suggests, such a process may well be one that the majority of European farmers cannot altogether win:

'Latest figures compiled by a team of international economists on an admittedly small sample of farms in the main dairying areas of the world suggest that even the best European dairy farms would have difficulty competing against the low-cost, grass-based systems of Oceania or even the largest and most efficient dairy units in the western United States. More immediately and importantly, however, these figures suggest that EU producers could not compete with the best central European producers on the

basis of current labour and land costs. In addition, the greater potential international competitiveness of the most efficient EU producers, such as those in the UK, could be eroded by currency appreciation' (Agra-Europe 1998 p. 2).

The globalisation logic of the *agro-industrial model* is reinforced further by partial de-regulation and the somewhat spurious faith in 'free' competition (for instance with abolition of the UK nationally based Milk Marketing Board in 1994, and subsequent fragmentation of the struggling producer cooperatives, on the basis of its monopolistic practices). In the meantime dairying is being practised by larger and more specialised units. In Wales, for instance, after long-term declines in the number of dairy farms in the 1960s and 1970s (down in numbers by 33 per cent between 1964 and 1974); the number of milk producers fell by just over 50 per cent from 1977 to 1996. This downward trend is set to continue.

While 'reform' of the CAP has been slow (either in developing a more effective rural development arm or in more radically reforming the price and production support structures) it is clear that the realities of the current agro-food conditions are still largely pulling it along an agro-industrial path which is based upon a neo-classical 'virtual' logic of scale and specialisation using industrial technologies. Such a model of development, however, is not without its growing threats. In Finland, for example, the number of farms is decreasing rapidly at the same time that average farm size increases. Since EU membership (1995), the number of farms has fallen by 3–5 per cent per year, and altogether by 18 per cent.

The agro-industrial model: the dynamic management of unsustainability

The industrial food governance system, associated with what Friedmann and McMichael (1989), inter alia, have called the 'first' and 'second' food regimes, owes its raison d'être to the tendency towards the reduction of distinctiveness of food production, supply and accumulation; and the alliance between food and other technological trajectories associated with industrial activity – where there is an in-built requirement to overcome the vagaries of nature and a rational approach towards increases in outputs. This, of course, has been the dominant model of food provision in the post-war period (see Buttel 1997), and has been conditioned in ways which attempt to stave off the unacceptable environmental consequences of such a system by developing new technologies (e.g. biotech and GMOs).

However, the direction and development of this model has not been without its tensions over the past two decades. It is worth illuminating these on the basis of identifying how the model has attempted to 'solve' – perhaps only over the short or medium term – some of its own problems

upon the agricultural alone. Rather, it is to the provision of rural services (both public and private) which can hold one of the significant keys. The growth of the 'consumption countryside' model also holds a particular conception of nature and the rural. While, for instance the 'agro-industrial model' sees rural nature as something to be overcome, or at least held back by the continuous application of new technologies in the agricultural and food sphere; the post-productivist model tends to marginalise nature in a different way. Here, rural nature is now a consumption good, to be exploited not by industrial capital but by the urban population (Enticott 2000). Moreover, it is usually a specific form of nature that needs to be constructed and exploited. It is centrally concerned with the construction of the rural landscape and its protection. This is seen as somewhat disembodied from agricultural developments, with producers now being seen as the potential 'criminals' in the moral crusade to protect these cherished rural spaces (see Lowe *et al.* 1997).

Underlying the 'post-productivist' conception of the countryside then is a pervasive 'moral economy' with respect to agriculture. The latter is seen as a 'dirty-business' to be cleaned up by progressive environmental legislation. In the mean-time a raft of non-agricultural policies can be more effectively directed at maintaining a rural economy and generating 'thriving rural places'. Such conceptions of the countryside are thus not just associated with the growth of the ex-urban middle class as the power-brokers in the countryside. They are also associated with a particular conception of rural nature; one which places most emphasis upon the protection of the visible landscape and the maintenance of the scarcity value of different types of rural environment.

We can see then that while the *agro-industrial model* tends to marginalise rural nature through the production process, *the post-productivist model* now does so through a highly materialist conception of the consumption process. Perhaps not coincidentally both sets of modernist conceptions and models owe much of their origin to *external controls over rural nature.* They also involve the development of external forms of governance over the rural in ways which attempt to position nature in a way which will be at least sustainable for some period of time in the future. Such conceptions are now embodied in government reports. For example, a recent Cabinet Office (1999) report in the UK argues from the outset that:

> '*Rising affluence and changing tastes have led to a reduced emphasis on the countryside as a resource for production and a greater emphasis on its potential for 'consumption', both as a place in which to live and in which to enjoy leisure and recreational opportunities.... More people see themselves as having a stake in the countryside (and in particular, in the state of its environment): and the views of rural residents have diversified, reflecting the changing social composition of the rural population*' (p. 29).

The consumption countryside is, however, a site of conflict. Continuing with our reference *point of the dairy sector,* we can see under this conception of development how the intensively organised dairy sector contravenes the growing moral economy of post-productionism. This is most clearly demonstrated in Lowe *et al.*'s study (1997) of dairy farmers and pollution in Devon, England. As they conclude:

> '*The politicisation of farm pollution led to the imposition of various controls on agriculture to restrain the excesses of agricultural productivism. What these environmental regulations meant in practice was that farmers were increasingly confronted by regulatory officials armed not only with new powers but also with a new moral authority. The exchanges that took place between them were not limited to what constituted sound agricultural practice, but touched on nature, morality and the law. Much of what was of interest to us not only in the regulation of polluting agricultural practices, but also in terms of the making and undoing of moral discourses, occurred in these encounters. What was revealed was a traditional order (of agricultural productivism) under challenge and a new order (around environmentally responsible agriculture) being discursively constructed and resisted*' (p. 207).

What we see here is the embodiment of conflict between productivist and post–productivist models being played out amongst the farm and wider rural population. A population that now has to confront not only the vast regulatory system associated with the agro-industrial model (as outlined above) but also the highly diverse and moral regulatory system stemming from a particular form of (externally imposed) environmental care. This form of environmental care or social management tends to criminalise dairy producers while protecting and shaping a countryside in ways which will remain particularly attractive to the maintenance of the counter-urbanisation process. Thus the *post-productivist* countryside model of development is founded on anything but environmental protection 'for its own sake.' It is far more functional than this. It is about socially and morally shaping the countryside in ways which continue to make it attractive and lucrative to aspiring ex-urban groups.

There are, of course, at the time of writing several types of rural backlash to these trends. In the UK the Countryside Alliance movement has grown rapidly into a force to project the anti-urban and pro-rural view of the countryside and in the US the anti-environmental movements (Jacobs 1999) project both productivist and various types of stewardship perspectives for land-owners and farmers. These conflicts occupy one contested territory concerning rural nature; a territory which brings together productivist with consumption concerns. Both, however, tend to marginalise and deny the critical role of sustainable agriculture in socially managing rural nature.

The sustainable rural development dynamic: pathways out of contradiction?

Both of the aforementioned rural development dynamics have their own socio-spatial expressions. In many rural regions in Europe they overlap across rural space and affect change in dual ways. For instance, the quiet progression of the industrial agro-food model (for instance in the intensively farmed lowland regions) tends in many rural spaces to continue as the post-productivist conditions also take hold. This creates more complexity about the ways in which rural nature is being governed; for instance, with externally managed CAP funds and quality conscious corporate retailers conditioning the actions of smaller groups of specialised producers, at the same time as creeping local and regional environmental protection and planning attempt to contain and shape the same rural landscape. The countryside becomes increasingly superimposed by more diverse forms of regulation.

The interesting contrast here is that the former (productivist) system of governance tends to stress the national and international perspective (largely ignoring local and regional specificity of circumstance). The latter, by contrast, governs by local means and places giving strong emphasis upon the specification of the local environment (see Kitchen 2000). Both, however – as I will explore in subsequent chapters – tend to down-play the potentially symbiotic social and environmental role of agriculture.

One important implication concerning the onset and dominance of both the agro-industrial model and the post-productivist model is that their own specific moral economies, there own specific ways in which social nature is managed, rely upon both market and state governance structures which attempt, over the medium term to manage the largely unsustainable conditions which they create (see Drummond and Marsden 1999; Drummond *et al.* 2000). These governance structures, whether associated with agro-industrial productivism, or differentiated types of locally-based post–productivism, maintain a tight grip in their attempts to avoid environmental and social crises of legitimacy. As Drummond *et al.* (2000) indicate this when referring to the growing crisis of European agriculture:

> *'Thus far, both the analysis and policy have failed to relate the progressive devaluation of nature to the reconstruction of regressive modes of social regulation, and debates about sectoral policy have progressed as if these interconnections and dialectics are inconsequential. It has generally been assumed that environmental and social policy can be bolted onto both agricultural policy and, more importantly, onto existing structures and relationships'. On the one hand policy is being articulated in terms of environmental and social goals, but because these are to be implemented through the existing farm structure, this may well serve to reinforce and*

> *reproduce the tendencies towards continued instability and environmental instability*' (p. 124; my emphasis).

These regulatory and governance mechanisms have a dual role in both upholding largely unsustainable agricultural practices and broader rural conditions at the same time as acting as *significant barriers* toward more radical overhauling of rural policy. In short, the productivist and post-productivist system of governance, while coming from different origins, have found that they can peacefully co-exist – at least for the time being.
Perhaps not surprisingly, therefore, is the fact that it has been in those regions which have been *least exploited by either model* which have given at least more initial impetus to alternatives: the *sustainable rural development model*. While it may owe its origins to those regions which have been largely 'passed by' by the other development logics (what are traditionally termed 'peripheral rural regions') it is now highly pertinent to more widespread rural contexts (see Table 1.1).
What are some of its emerging features in relation to agro-food? And to what extent can we begin to see alternative and robust regulatory and governance structures taking hold? Quite significantly we can begin to see a broader definition of nature which now encompasses not just its surface or landscape value as a consumption good, but one which gives more emphasis to food production and the natural transformation process in itself. Also the social and economic aspects of sustainable development are more clearly to the fore. So to what extent does this represent a new departure for rural governance of nature?
In variable ways this alternative rural development model conforms more effectively to Guzman and Woodgate's definition of agro-ecology (1999) which encompasses:

> '*the ecological management of biological systems through collective forms of social action, which pursue co-evolutionary pathways that address the needs of society/nature without jeopardising the integrity of society/nature. This is to be achieved by systemic strategies that promote the development of more equitable and flexible forces and relations of production, and modes of consumption and distribution. Central to such strategies is the local dimension where we encounter endogenous potential encoded within landscapes and knowledge systems (local, peasant or indigenous) that demonstrate and promote both ecological and cultural diversity. Such diversity should form the starting point for alternative agriculture and for the establishment of dynamic yet sustainable rural societies*' (p. 26).

While this is an over-arching and ideal-typical type of definition, there is now sufficient evidence across Europe that many of these components are beginning to emerge, largely it would seem, in contrast to our earlier and more dominant models, at a regional, rather than at a national or specifically local level (see Marsden *et al.* 1999).

This is marked by a different set of organising principles which place nature, labour, region, value, and quality in a different set of equations. Not only are the foods themselves representative of different quality conventions (i.e. ecological, spatial, artisanal) but the mode of social organisation of supply chains is radically re-organised. The connectedness of supply chains is related to new types of association, and the types of innovation are linked much more to organisational as opposed to technical developments. The factors of production are reconfigured in ways which place capital in a more symmetrical relationship with labour and land, and emphasis is placed upon the creation of positive externalities from a more multi-functional land-based production base. The emphasis is upon diversity and divergence rather than convergence and consistency. Production and supply chain forms are more highly variable and 'context-dependent.' Value is captured more effectively at the producer end of the chain and the use of spatial identifiers for the foods produced becomes an important dimension. (see Table 1.1)

Such an 'alternative system' requires new forms of quality regulation to legitimise its presence. Outside the EU designated categories (e.g. PDOs etc) and various certification bodies associated with organics and green products, these tend to (i) be more recent, (ii) be locally defined on the basis of the production unit or the locality; and (iii) more associated with satisfying the composite character of the consumer concerns mentioned above. These regulatory systems have not simply developed 'top down'; and national government policy and regulation is currently only marginal to the governance of the system. Indeed, because of the diversity and the emphasis upon place, nature and quality, national governments find it increasingly difficult to control and regulate the alternative system. This is one of its main features, and it represents a major constraint upon the widespread development of these systems. A major reason why the rational and industrial systems of regulation are often seen as more attractive and legitimate from a government point of view is that they are more 'controllable' and therefore more conducive for adoption, particularly under conditions of growing crisis.

In recent years, a number of alternative approaches to food governance have emerged, their development coinciding to the rapid growth in influence and power of the multiple retailers. The emergence of 'forces of opposition' to the dominant model of food provision raises interesting questions, especially at a time when the hegemony of governmental influence over food policy is in crisis. The BSE crisis, and to a lesser extent, the debate over GMO foods have done much to undermine the public view of the legitimacy and impartiality of the state in situations where agro-industrial and consumer interests collide (ESRC 1999; Marsden *et al.* 2000b).

Against this new background new forms of food governance have been emerging on an ad hoc and grassroots basis. In the past five or so years they have acquired a 'critical mass', challenging both commercially and ideologically the dominant modes of governance. What are some of the patterns of development in these alternative food networks?

Product focused alternative food

Of the new modes of food governance being constructed at present the most significant, for a number of reasons, is that of organic produce. Although not 'new' the conscious adoption of organic standards by increasing numbers of farmers and consumers represents a potentially significant challenge to the industrialisation and intensification of agri-food systems, representing a desire to return to more natural values. This would not have come about if not for the growing concerns over the safety, health and environmental and animal welfare implications of the existing system, all of which are implicitly challenged by the rigorous standards – now defined by EU legislation (Council Regulation (EEC) no. 2092/91, OJ L 198, 22.07.1991, pp. 1–15) – but open to slight variations in interpretation at the national level) which ban the use of artificial fertilisers, pesticides and pay careful attention to animal welfare. Consumer demand for such products continues to rise and far outstrips supply, especially in northern Europe (Michelsen *et al.* 1999).

The emergence and growth of thriving organic sectors across Europe poses challenges to the dominant system of food governance, but also offers new opportunities. On the production side it implies withdrawal from the agrochemical complex, greater reliance on local sources of fertility, a reversal of the fifty year process of intensification and specialisation. Yet it also offers opportunities to tap into new markets at a time when commodity prices, for example for milk, have sunk to a fifteen year low. This is especially relevant for marginal regions in both northern and southern Europe, which cannot realistically expect to compete with the new super-productivist, global regimes bought about by the new CAP and WTO agreements.

After more than a decade of stop/start policies towards organic produce the multiple retailers are showing signs of serious commitment to promoting and developing the sector. However, engaging with the organic market is not entirely a risk free strategy for the multiples. For with organic produce more widely available on the shelves of supermarkets, consumers may be given more reason to question the provenance of 'non-organic' produce. As a recent report from the (Danish) Ecology Board points out

> *'In today's environmental reality consumers demand organic milk in order to avoid milk which contains pesticides. In the old days this was simply called milk. And so it should be again. Until then the alternative should be labelled*

pesticide milk. Just as supermarket shelves should contain eggs and battery eggs' (in Halkier 1999).

Regional, local, and artisanal foodstuffs

A parallel trend in production and consumption patterns is the emergence of a strong and growing market for regional, local and artisanal foodstuffs. This represents a major discontinuity from the assumed dominant trend within the food industry for centralisation, standardisation and increased homogeneity. In many respects such products represent a manifestation of the 'small is beautiful' movement. Their defining characteristics are small scale of production, using high grade raw materials (often but not necessarily of local provenance) and artisanal/traditional production techniques to create quality items which attract premium prices which for the most part are sold through local and/or specialist shops rather than large retailers.

Interest in such products stems from a number of sources including the rapid growth of a revised 'food culture' and increasing concerns over the safety of mass produced foodstuffs (Nygard and Storstad 1998). The growth of the domestic market for such products is rapidly growing in the UK, according to one survey by 20 per cent over the past three years and now accounting for 5 per cent of grocery expenditure. (DTZ Pieda Consulting 1999).

Some products are genuinely rooted in tradition (e.g. Scotch Oatmeal cakes) others may have long traditions of production but have only been recently re-branded themselves in order to differentiate themselves from similar products (e.g. Welsh Black Beef) and some may be relatively new product lines which, nonetheless embody quality and authenticity. Unlike their equivalents in southern Europe few British artisanal foodstuffs have, as yet, been accredited under the PDO/PGI system.

In Ireland, an interesting battle has developed between the *agro-industrial model* (as applied to dairying) and alternative rural development model solutions. The latter has been adopted in such schemes as the County Clare Partnership Dairy Action Research programme, designed to add value and to diversify the dairy product base for the smaller dairy farmers (Kinsella *et al.* 2000). Also, as McDonagh and Commins (1999) indicate, the growth of what they term the 'speciality sector' as part of the Irish rural renaissance, even if it is not completely clear whether this will lead to dual strategies of agro-industrial incorporation or continued independent adaptation. Nevertheless, what we witness here in these examples in the Irish dairy sector (as well as some examples in the Dutch dairy sector). are counter-productivity strategies as part of a greater rural development paradigm. These focus upon 'economies of scope' (Saccomandi 1998) rather than scale. The case of dairying as an exemplar here (as followed through with examples from the agro-industrial, post-productivist and rural development models alone) demonstrates the different social management and governance of a natural rural product. The examples also demonstrate that the dominance of the agro-industrial 'race to the bottom' scenario for the dairy producer sector is by no means inevitable; it just has to be socially managed in ways which heighten the significance of milk products, the land base and the producers as social/natural actors in food governance systems.

Interestingly, these regionally identified products have emerged at a time when there is an increased emphasis on the broader cultural assertion of regional and local identities. They also emerge at a time of economic crisis in farming, when the mantra 'compete on quality rather than price' is central to many governments rhetoric on the future of agriculture. It is recognised that 90 per cent of such companies are rural (MAFF 1999) in the UK, and play a growing role in the economy of the countryside. Their future growth, and the propagation of new businesses are likely to be key planks in the forthcoming discussions about the future of the regional rural economy.

Direct food supply chains

One important aspect of the centralisation and standardisation of the industrial food command system developed by the multiple retailers include requirements for continuous supplies of high volumes of standardised produce. These requirements have effectively excluded artisanal, regional and speciality food stuffs, and in many cases organic produce from the dominant food chain.

In reaction to the difficulties faced by small scale producers in meeting the standards (whether of quantity or quality) a number of new initiatives based around developing more *direct, often local based supply chains* are emerging. These include: box delivering schemes, developing closer links to local retail and catering trades, farmers markets (Festing 1998); the use of the internet and mail order (Mai and Ness 1998), farm shop sales, and at the same time a suspected increase in rural areas in the black market of farm-gate meat sales. There has also been a marked increase (after some fifty years of decline) in urban food production, with a growth of interest in the most direct form of food supply chain: self provisioning (Martin and Marsden 1999).

It is possible that the development of these new 'short supply chains' represents a more profound shift in governance patterns than the emergence of new products, in that they are being explicitly used to create markets independently of the industrialised and intensive ones led by the globalised multiple retailers. Markets are being forged with new forms of social and association relationships, where alternate quality standards, generally favouring the ecological and the artisanal come to the fore, and where there is a reassertion of the value of the local, and whereby producers, or those closely linked to them retain a greater proportion of the value added. Many of these initiatives have been led by organic producers frustrated at the difficulties of working within the dominant food supply system.

In the development of these alternative types of food networks, the relationship between food production and distribution on the one hand, and environmental sustainability, human health and regional development are made more explicit, with producers, retailers and consumers apparently aware of a broader range of values other than the quality (defined in cosmetic terms) or the cost of food itself. The provision and consumption of food provides a link to issues encompassing health, nutrition, the environment, region, community and third world development. The divide between production and consumption is bridged by direct producer-consumer contacts and the renewed seasonality of produce, now virtually abolished by continuous availability of most lines in the agro-industrial system. More detailed empirical analysis of these alternatives forms a central focus of Section 3 of the book.

Conclusions

This chapter has outlined three ideal types. They are three pervasive dynamics which are shaping the governance of nature and European rural space in contested ways. They stem from different origins and they owe their existence to sets of different social, economic, political and academic/scientific constellations. The *agro-industrial model*, for instance, is upheld and reinforced by many agricultural economists; many of which are still bemused by the scale and technological expressions imported from the American agricultural modernisation project. The *post-productivist model* is alternatively associated with the social geographers' and planners concerns of rural capacity and the coping mechanisms needed for dealing with the 'consumption countryside.' This stems from the growing 'hegemony' of what Keil (2000) has called the 'suburban frontier', whereby, rural resources and landscapes are increasingly shaped by an extending suburban culture which needs to redefine itself with nature.

The rural development dynamic, while originally having its origins among the largely 'bottom up' initiatives associated with integrating and empowering 'peripheral' rural communities, (most notably the LEADER initiatives), is now a much broader and diffuse church; one which can incorporate renewed ideas of former agricultural practices and social ecology (see Light 1998). All three of these are, in themselves, types of modernisation project. The question becomes which will develop a new dominance and how will they play themselves out over the differentiated rural space?

The exploration of these three dynamics, and particularly the development of the emerging but so far marginal rural development dynamic, raises some important questions concerning the *regulation and governance of nature and rural space.* The rural development model demands, albeit in variable form, that nature, region, quality and value are reconfigured around the character and shape of rural resources (as the Guzman and Woodgate (1999) reference above implies). Of course, as the post-productivist model exposes, rural areas are more than the sum of their agricultural parts, but we may now see pathways for bringing agriculture 'back in' as a significant manager of rural nature, both for productive and consumptive purposes.

Of critical importance, if the renewed rural development dynamic is to take more of a hold, is the degree to which it can assemble legitimate governance and regulatory structures and processes which are integrative and robust. These will, it would appear, have to also more spatialised. It is only at the local and regional level that real socio-cultural and environmental integration can occur, such that 'higher' forms of authority (say from the reform of CAP or national government interventions) may be capable of setting overall strategy and 'pipe-line' funding mechanisms, but not so capable of carrying these through in an integrated and deliverable way. This is very much the spirit, if not the complete purpose of the reform of the CAP and the creation of the of the 'Second Pillar' as well as its associated regional and national rural development plans. However, such mutli-level hierarchies need to be 'cut through' by a series of integrating mechanisms associated with alternative governance approaches. As Lipshutz (1999) argues more broadly in relation to environmental governance :

> *'Much of the implementation and regulation inherent in regimes will have to take place at the regional and local levels, in the places where people live, not where the laws are made... Such governance is characteristic of the emerging global political economy characterised by economic integration and political fragmentation. The fundamental units of governance are, in this system, defined by both function and social meanings, anchored to particular places, but linked globally through networks of knowledge-based relations. These relations develop when the costs of acquiring information through 'normal' channels becomes too great. These relations bear a remarkable resemblance to transactions and economies oriented around kinship relations-and bio-regionalism too – in which trust and membership replace formal hierarchies and markets.. The phenomenon of 'networking,' characteristic of relations within global civil society, resembles this form of organisation. It is a form that lends itself to cooperation without centralisation, without 'global management'* (p. 113).

The alternative food chains outlined here, for instance, can be seen as representing one emerging manifestation of such an alternative governance of nature. Conventional producers are increasingly forced into such arrangements in some regions because of the high transaction costs operating in the conventional chains (i.e. associated with hygiene and food quality regulations as much as with the actual cost of food transaction). Of course, at least so far these emerging new systems of governance are far from dominant or comprehensive; but they are developing, albeit in variable form in different rural regions of Europe.

However optimistic we may be, concerning the renewed certainties such a rural development model may bring, or indeed, excellent exemplars it already portrays; it has to be recognised that its very development confronts many of the barriers associated with the established governance and regulatory systems of the other development models. At one level, for instance, the rural development initiatives associated with CAP reform would suggest whole-hearted support for the third model. But its fledgling birth coupled with its differential adoption by member states demonstrates the difficulties such a model has in competing with the logics of agro-industrialism and post-productivism. Everybody agrees that the CAP needs reforming, but many member states (including the UK) are tending to hide behind a duality/contradiction. Reform means movement towards the globalised agro-industrial model on the one hand, and at seemingly the same time, towards at least a gesture towards the rural development model on the other.

What is important to recognise in amongst this seeming 'reform' is the maintenance of the inherent conflict between the agro-industrial model and the rural development model. Most governments will give their blessing to the development of 'value-adding quality food production'; but they do so with their other hand firmly shaking the super-productivist hands of global agribusiness; and all that this implies for the continuance of the productivist agricultural treadmill. This can only act to provide barriers to sustainable rural development through marginalising its impacts.

In conclusion, therefore, in governance and regulatory terms we cannot expect, it would seem, national governments or corporate firms to fully stimulate the rural development model. *Its impetus will have to come from building alliances from within – at least at the regional and local levels.* The governance of rural space, however, is likely to become even more – however reluctantly – drawn into the re-definition and valuation in nature. This will take it far beyond the simple exploitative *modus operandi* of the agro-industrial or post-productivist models of rural development. The big question is how can emerging and durable regulatory and governance frameworks be developed so that existing sustainable rural development exemplars turn into common trends and practices across rural Europe?

We have seen here, therefore, how, on the one hand, we now witness an agro-food industrial system which is prone to increasing crisis; but an emerging more sustainable rural development paradigm which is still held back in its wake. The development of a new agro-food governance, one which plays a fuller part in assisting the achievement of sustaining rural nature becomes more and more urgent. The subsequent three sections of the book explore the three competing models and dynamics outlined here in more theoretical and empirical depth. The aim to demonstrate the critical role of social-science analysis in exploring the barriers and opportunities involved in progressing the condition of rural sustainability. Omplicit in this analysis is the argument that in only exploring criically all three and dynamics can realistic alternatives be sensitively postulated. As a result, rural social science has a central role to play in both critical and normative analyses in rural development.

The following two chapters explore the agro-industrial model in both the south (Brazil and the Caribbean) and in the north (Wales). They demonstrate the contradictory, if pervasive, embeddedness of the agro-industrial paradigm.

Section 1

The Agro-Industrial Dynamic

2 Global food spaces: understanding the uneven agro-industrial spaces

Introduction

The 20th century exhibited a process of residualisation of the agricultural and the rural. Food has been taken for granted. This residualisation process has now come to an end. For a variety of new reasons (diet, health, amenity, environment-nature concerns, reactions to modernism) the rural and the food resource defines a new centrality in peoples lives. Its value is in a process of differential reconstruction.

This chapter attempts to set out some revised concepts which give more clarity to the 'post-cold-war' development of agrarian space and the significance, more broadly, of food. The analysis is based on recent research conducted in Europe, Brazil and the Caribbean, and specific reference is made to some of the findings in the second part of the chapter. First, it is necessary to make some general points on the critical standpoint. In previous papers, I and other colleagues (see Arce and Marsden 1993; Marsden and Arce 1995) have been reconstructing an analysis of globalisation and food which provides a more sociological and geographical perspective, incorporating social agency, contingency and uneven development into analysis. This has been done in a spirit of building upon the more macro/food regime models, and has focused specifically in attempting to understand and conceptualise the new patterns of globalisation.

In the 1990s there are new agrarian questions to be addressed, requiring a more sensitive analysis which considers the expression of contradictions and contingencies of post-war capitalist development. In this regard there is a need to move from a broad conceptualisation of *food regimes* to one which focuses upon *food networks*. In this chapter I will attempt to justify why it is necessary to reconsider earlier conceptions, and begin to point to the ways in which this can be done. The significance of this task is heightened in the contemporary period. Food networks play an increasing role in the social and political development of regions and nation states, (i.e. it' s not just 'global'), in new and uneven ways. First, it is necessary here to outline some points of departure which set the scene for the development of food network approaches.

Some recent departures: food networks and value

It is important to recognise some key changes in the dynamics of agro-food networks.

(i) Aspects of control, power and dependency in agro-food networks are not only based upon input-oriented corporate capital (i.e. agribusiness firms and their relationships to the farm-based sector). Increasingly, they are associated with the control and construction of *value* from the point of production. This serves to empower near-consumer agencies. This reflects the common point that 'value adding' in food has tended to be associated downstream of the point where it is produced.

(ii) Food networks are characterised by the significance of 'action at a distance', as well as with local and regional 'nested' links, whereby the design of foods are shaped largely by the non-farm sector.

(iii) Globalisation of food is paralleled by intended and unintended local and regional effects, such as the reconstitution of local diets, food scarcity in potentially agriculturally abundant places, and surges of over-supply during volatile international currency conditions. These, in their turn, shape globalisation processes.

(iv) Technological developments and their social effects (in terms of our established concepts of subsumption, substitutionism and commoditisation) need to be situated in the broader context of knowledge networks; particularly surrounding the ways in which food quality is constructed. We are moving towards food supply chains which are qualitatively regulated rather than quantitatively. As a result, food 'value' becomes increasingly multi-layered, containing both commodified and formerly non-commodified attributes.

(v) Nation state regulation does not always have a functional relationship with aspects of globalisation. Global processes operate largely on the basis of corporate manufacturing and retail capital; but national processes still have to maintain 'a public interest' for legitimatory reasons. Corporate retailers (and their relationships with the nation-state) become dominant actors in the reconstruction of internationalised and national food networks (see Marsden and Wrigley 1995; Marsden and Wrigley 1996).

(vi) Food markets are becoming 're-regulated' around the demands of segments of consumers (in the North, and in the 'Westernised' parts of the South) and corporate retailers. This involves the 're-naturalisation' of foods as well as their continued industrialisation.

(vii) These processes reproduce different mosaics of regional production and consumption spaces, new *dynamic* and *dependent* spaces providing major forces for redefining uneven development.

Three such global/regional spaces include: new agro-industrial districts (such as, Sao Francisco Valley, Brazil); quality consumption and retail spaces (much of Northern Europe); agricultural retreat and local vulnerability spaces (the Caribbean). These spaces are interlinked and cross-cut with dependent relationships and networks of power. For instance, under globalised conditions the legitimatory relationships established in one place (Northern Europe) may be significantly at odds with those in another (*e.g.* the Caribbean). As a result, and if food commodities move between these places, they may carry their natural features (freshness, colour) with them but leave their more embedded social legitmatory relations very much behind. Indeed, it would seem that, valued consumption spaces need to continuously devalue other spaces in order to reproduce more sensitive capital accumulation. Distances have to be reproduced even, indeed particularly, in a more deregulated global terrain. In this sense globalisation and deregulation tend to redistribute risks and power, making it easier in one place to extract value only at the expense of other places. This becomes a key aspect of uneven development – the parallel but contradictory construction of value – under the new globalised conditions. In this sense, globalisation is anything but a new surface eco-geometry. Rather, as this chapter attempts to show in comparative focus, it is socially insinuated and embedded into the regional organisation of food production and consumption.

These sets of principles emerge from a series of dialectical relationships and contradictions in the agro-industrial food sector. These concern: production/consumption; globalisation/localisation; deregulation/re-regulation; valuation/de-valuation; powerful spaces/dependent spaces; cultural distinctiveness/social exclusivity. It follows that analysis of food and rural resources in the late 20th Century needs to explore how these contradictions are sourced, spatialised and played out, and how key actors and agencies are engaged in them. This places an emphasis not only on systemic approaches. It requires also a focus on how discontinuities and contradictions are often *put together* in order to make things happen.

Quality, regulation and consumption: key concepts in the reshaping of the new industrial agrarian spaces

Value-adding has become the central concern for the post- farm parts of the food chain (Rabobank Nederland 1994). However, to concentrate on the increasingly diverse ways in which corporate capital achieves this through the redesign of foods tends to restrict our focus. The construction of value of foods, and of the labour and environmental conditions to produce and convey them, requires a renewed and more integrative focus. The construction of food quality has to be seen in *the context in which it is embedded in environmental value and labour value*. With this in mind it becomes important to more broadly consider social theories of value and the broader definition of commoditisation. Regulation and regionalisation are inputs and outcomes of these social processes.

In empirical work in Brazil and the Caribbean (referred to in both this chapter and Chapter 3) a focus on this value construction in the fresh fruit and vegetable sectors has demonstrated that the traditional constraints on land-based production contribute only one element in the distinctiveness of agrarian development. The value of labour, and the environment are traded-off against the priorities for particular conceptions of food value and quality. Risks are allocated in the networks of food supply (e.g. producers, exporters, traders, regulators) around the need to maintain different quality criteria. Under these conditions powers and risks become rearranged around actors engaged in the carrying and exchange of food goods. The externalisation of markets empowers some of the carriers and exchange actors over producers, and indeed over some national state regulators. Some networks are formally organised, some are not. Both have to deal with uncertainties translated from distant consumption spaces.

In the dominant consumption spaces in Europe, corporate retailers play a key role in translating the quality definitions of foods globally. They become key players in the allocation of risks and constraints in supply networks. In addition, they develop their own regulatory systems which ensure a dominance over the supply of main food products. This is a highly competitive and contingent process which is vulnerable to consumer reaction. Its success depends upon the social and political maintenance of a set of power relations which have to be legitimised. In consumption spaces of the North, this is increasingly an exchange-based system, rather than one dominated by industrial relationships. 69 per cent of European consumers claim to most often buy their groceries from supermarkets, with consumers making an average of 2.14 supermarket shopping trips per week (Tordjman 1995). Moreover, on the supply side, the saturation of the quantity of foodstuffs in most developed markets has resulted in a more or less constant or declining value for the primary sector, despite rising consumer expenditure. In the US, for instance, 'farm value' has remained virtually unchanged in recent years whilst 'the marketing bill' by 1988 was almost twice what it had been a decade before. The marketing bill – incorporating the added value of the industry, trade, retail and food services – is three times as large as farm value.

Some analytical implications

Some of the principles set out here justify the need to counterpoise micro-sociology with macro concerns. Both need to be brought together in new ways. This requires conceptual development whereby social actions can be explored in their different contexts of value construction, associated with globalised and regionalised food networks. New forms of institutionalisation, regulation and spatialisation become significant in the uneven development of agrarian space (Watts 1996). This involves exploring differential processes of social translation. Global processes have actual translations at different spatial and social levels of interaction. They are constantly internalised and layered by different actors in networks of relationships. In particular, the bases of social action in globalised food networks are associated with quality, regulation and consumption. However, to recognise this does not explain how globalised food networks come into being; how they may be perpetuated or adapted; or what the differing local and regional effects of these transactions and relationships might be. In addition, we are unclear how these concerns lead to new forms of local and political consciousness which can then act to reshape local and regional spaces. In positing these questions in a rural context, we have to begin to look at both *the people* and *the foods* in the networks of supply and how these share or allocate power and responsibility, re-locate risk and penalties; and begin to construct international markets and uneven forms of regional development. As Feierman (1990, p. 36) argues more generally:

> *'The wider world is not external to the local community, it is at the heart of the community's internal processes of differentiation.'*

As we shall see from the case examples, a focus upon the ways in which globalisation becomes incorporated into the social life and action of industrialised food networks – in different spatial formations – raises important questions concerning the ways that food networks themselves are situated and embedded in localities. With reference to the case studies below, we can begin to illustrate how the relative significance of food and food networks in and through spaces is variable, according to the main activities in those networks, and the relative power to capture the social value of food products.

In addition, and as we have argued elsewhere (see Marsden *et al.* 1993; Murdoch and Marsden 1994; Section 2 of this book) food networks as a whole are cross cut by 'lateral' processes of rural restructuring which have little to do with the food sub-sector. Nevertheless, these boundaries of economic, social and political activity are in a process of transformation and *reconnection* in both the North and the South. A focus upon the situational position of food networks in different spaces begins therefore to raise important questions concerning the elements involved in the processes of uneven development. Food networks thus hold vertical and horizontal connections with the spaces in which they are situated. The social sums of both begin to reshape rural space both 'from the inside' and externally *with other spaces*. Socially and politically defined dynamisms and dependencies are formed and reinforced. A key spatial issue of concern here is that dependent, as well as dominant spaces are created.

In what follows we can illustrate these features which are creating uneven agrarian spaces by reference to different *but related* case examples. These provide an initial basis for developing a more effective comparative analysis of globalisation and regulation, focussing specifically upon the differential and spatialised construction of value of foods.

The development of new agrarian districts: the Sao Francisco Valley, Brazil

The Sao Francisco Valley development represents the largest irrigated agricultural development in Latin America (Figure 2.1). It is based upon the development of export agriculture, particularly mangoes, grapes and tomatoes. By the 1990s, the realities of significantly increasing exports, partly through the expansion of irrigated lands, but more recently through productivity increases, have to be underpinned and conditioned by constructing and maintaining specific and situated quality conditions of the fruits (i.e. mango, tomato, grapes and more recently, acerola). The nature of these quality designs and conditions extend from planting through to the point of consumption. They are not simply associated with achieving the criteria set out by either the domestic Ministry of Agriculture (Embrapa) nor those of importing nations and retailers (such as the USDA). Even though these agencies do have increasingly strict criteria, involving site inspections and continuous monitoring, it is important to recognise that surrounding these legal and formal values lie a series of conditions which are set by other local and external agencies. While the origin of these primarily deals with the dynamic maintenance of the quality of the food product, this has to deal with setting parameters for the specific labour and environmental conditions as well. The construction of these quality conditions, their implementation, and their ability to allocate risks and responsibilities to different parts of the food network, means that they play a significant role in constructing and maintaining the fruit supply networks themselves. In the highly regional and local competitive conditions, where overall aggregate supply becomes less of a concern to importing nations than those concerning specific design and quality, producers and exporters have to be geared to providing a consistent 'quality product'. This may vary between importing countries and it demands different production, packing and quality control systems.

Figure 2.1 Location of the São Francisco valley, Brazil

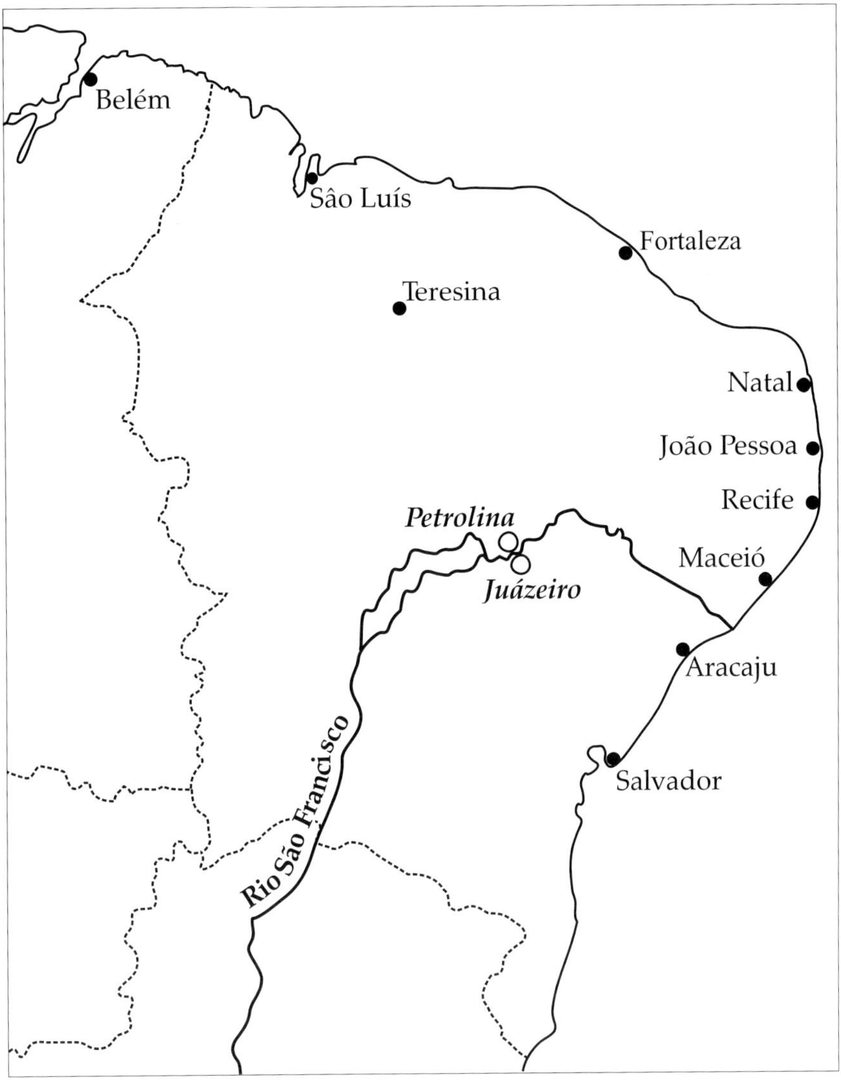

This concern for quality and design increasingly emanates from the strategies of the corporate retailers situated at the consumption end of internationalised food networks. For instance, in a rather economistic and aggregated way a trend report on the global food trade argued:

'The retail chain makes strong demands on security of delivery and product quality. The need for large scale and security of delivery leads retail chains into backward integration. This is becoming more easy for the retail chain to effect through the increased scale of supply and the liberalisation of world

> *trade. This may lead to the exclusion of the importer. Exclusion takes place earlier in the case of simple products like citrus. The importer will have to continue to manifest his added value for the retail chains. In addition to supply and guarantee of quality and logistics, storage and transhipment are important functions, through which delivery can be on demand. The transport function can be partly contracted out*' (Rabobank Nederland 1993).

At the local level those actors and agencies who attempt to control these diverse quality forms begin to play a pivotal 'social carrier' role in the agro-industrial region. Petrolina, the main urban service centre, becomes a focus for merchants, agronomy, trade and communications which act to solidify and extend local producer and exporter networks. An agro-industrial district begins to be formed. Some of these key actors and agencies become the conduits of globalised knowledges, and internally, the gatekeepers to globalised market entry for the numerous and variable producer sector. Agreements and contracts have to be socially based and continually reviewed. This places a considerable pressure on the maintenance of quality management and its articulation through food networks.

At the heart of these processes of food network construction is the differential creation of value through the coming together of different actors to produce and shape nature. This produces a major social process. In the Valley, the gradual divestment of state responsibility for the agricultural development based on irrigation (under CODEVASF) over recent years has led to the development of large integrated agricultural production enterprises that have developed an export function. Fruit exports can be sold through farm co-operative organisations (such as VALEXPORT), but the increasing scale of development in the region has spawned new private enterprises which focus specifically on promoting and gaining entry to export markets (e.g. FRUITFORT, MAPEL). While most of these have benefited from state supported development projects (e.g. NILO COELHO) in terms of infrastructure and initial technical assistance, they are effectively privatised firms which control extensive areas of production and recruit other smaller producers in order to export their products. These firms are located in pivotal positions in the globalised food networks. They can place strict quality control criteria on smaller family producers, and they can balance the proportion of fruit which comes from them vis-à-vis their own estates. In addition, because of their modern packing house facilities they are well placed to translate globalised knowledges concerning retailer and consumer demands concerning specific colour, size, shape and content of the crops. This information is translated to local producers who may have agreed a temporary export contract with the exporter enterprise. As a representative from one of the largest export firms (with control of 1,015 has of irrigated land largely devoted to mangoes and grapes) argued:

'Quality control is very severe. We have to follow the product process. We go onto the farm and inspect regularly. Nevertheless, possibilities are wonderful. Conditions are good. I cannot see any bottlenecks to this production. We have the most qualified and technological status. In Sao Paulo they are being eradicated.'

This company was currently arranging an export programme to Japan, with most of current production going to the US and Europe. By 1994 it exported 1 million boxes of fruit. By 1998 its volume increased to 2.5 million.

As well as the degree of 'arms-length' control these enterprises place on their enrolled producers, they have to take particular care in the harvesting and handling of the fruit from the trees to the packing houses. Also, the timing of this process has to be modulated to fit the gaps in market supply of the importing countries and the maturation of the fruit such that it reaches the consumers table in the 'correct' condition. This distinctive process of time synchronisation not only to market entry but also in the managed maturation and quality of the fruit is crucial to the continuation of the food network itself. For producers and exporters alike, a particularly vulnerable *time-quality episode* is that concerning the multiple and embedded handling and packaging of the fruit. The quality and its risks can change with the mobility and handling of the product. Also, large differences can emerge between family producers and large enterprises, leading to the rejection of fruits for the market. One exporting firm had ceased taking the fruits of small producers because of this variability in control and the effects it had on the delicate appearance of the mango. It is necessary to achieve an estimated time of 23 days between harvest and actual consumption of the mango. The appearance and quality can severely decline after thirty days. Most of the exporters are responsible for the fruits until they are landed in the host country. Also, they may not know who the buyers are until they have left Brazilian shores. Dealing with these global contingencies but at the same time maintaining quality surveillance of the products is the responsibility and major concern of the export enterprise. Moreover, the precise links in the networks are not necessarily complete or clear during critical time-quality episodes. The whole network does not always exist at the one time. Rather, different time-quality episodes are progressed from farm production, through harvesting, packaging, transporting and distributing products. This contingency, rather than being a random or residual event, becomes a major social context within which risks and responsibilities are allocated.

In the process production, harvesting, packaging, the inevitable handling and carrying of products comes with its own risks. This implicates and promotes a gendered set of labour relations (see Chapter 3). Within the protocols concerning the limits on pesticide and herbicide uses the problems of the biological risks to the fruits is a real one. American importers, for instance have become concerned with the incidence of the Fruit-fly. This has resulted in a more stringent quality control procedure being instituted in the local packing houses. This involves the drenching of mangoes in water and petroleum bees-wax; with the grading process being followed by USDA officials who are rotated at monthly intervals. They have to be paid for by the exporter firm. For the European market and packing process there is a more rudimentary process, with little checking or external monitoring. It is argued by producers and exporters that drenching changes the inherent quality of the mango, but that the effect of communicating and signifying safety, standardisation and aesthetic pleasure (by adding an externalised value to the product) ensures the maintenance of the American market.

This example of how one of the importer countries' concerns and conventions for quality get *fixed* into the operation of the network demonstrates how quality parameters and fruits and their labour processes are malleable to different exporter and importer concerns. The constraints inherent in the disparities in production and labour time in agriculture *vis-à-vis* industry (see Mann and Dickinson 1980; Mooney 1985.) are only one set of social constraints which give fruit culture a distinctive position in modern capitalist agrarian development. The malleability and the sensitivity of many of the products (in this case particularly grapes and mangoes) coupled with the growing external demands for specific quality conditions, means that the management in the food networks of the specific time- quality episodes from harvesting to point-of-sale provide a significant and distinctive basis for social action and the reconstruction of value in food networks. In such regions as Sao Francisco, where irrigated systems can establish a harvesting pattern which continues for eight months of the year at least, this constant process, of managing and co-ordinating time-quality episodes is a major and dominant feature of the agro-industrial region and the networks of food which flow through it. Those actors and agencies who are closest to the definition and implementation of quality conditions begin to accumulate power in the food network. This leads to a growing social and economic differentiation in the region, with smaller producers, not exclusively, but prone to exclusion in globalised food networks.

Global and regional dimensions: regulating vulnerability in the Caribbean

Agriculture and environmental resource use in the Caribbean and in Barbados specifically has to be seen within the broader structural context which partly defines it (Figure 2.2). It is necessary to realise that the power to define resource use predominantly lies with the logics of global forces even if these can be adjusted and resisted at the local level. Of particular significance in understanding the variable deterioration of environmental conditions in many of the small islands Caribbean (Watts 1996), is the need to appreciate how agriculture is both externally and internally defined by particularly powerful actors and agencies.

Figure 2.2 Location of study islands

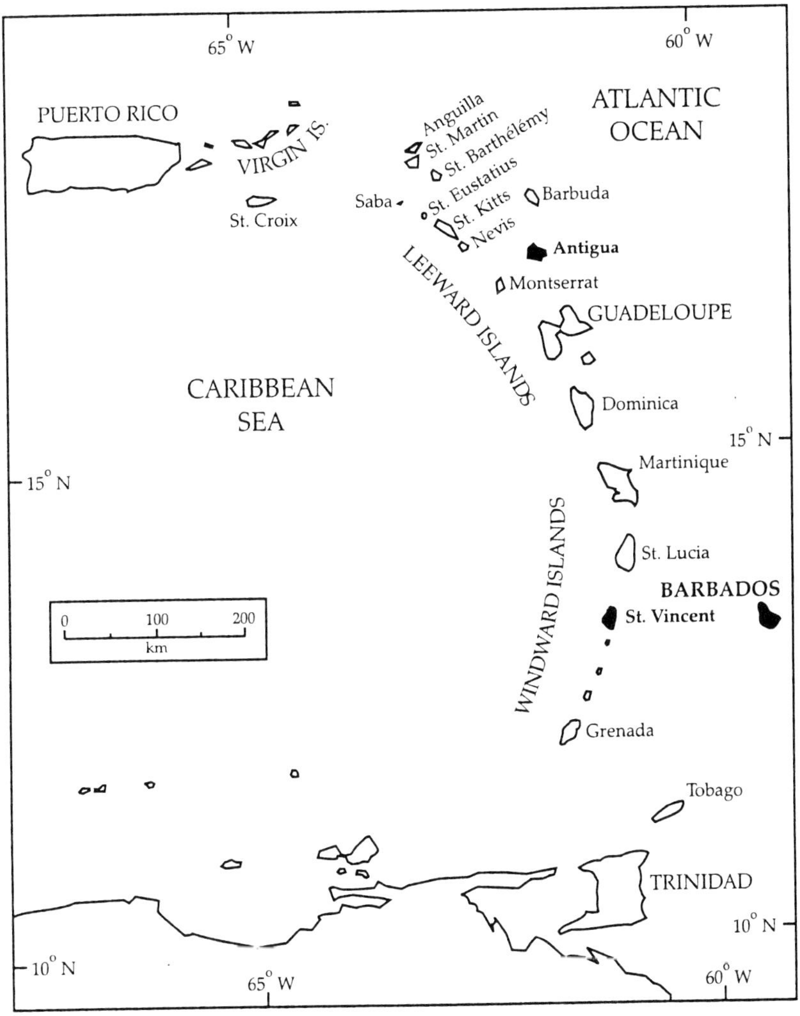

Largely, this definition sees agriculture as a predominantly export commodity system capable, to varying degrees, of generating foreign exchange and easing nationally based debts and balance of payments conditions. This logic defines notions of efficiency on the one hand and marginality on the other. Small farmers, once seen as marginal by the plantocracy are now marginal to the external commodity markets. Agriculture is largely stripped of its ecological and human significance (see Polanyi 1944; Fitzsimmons 1989). Under these conditions environmental vulnerability also becomes marginal to the priorities of production. As a result, and despite serious efforts to improve agricultural systems, this region is characterised by the progressive devaluation of food and agriculture.

In this context it also has to be recognised that agriculture and food have a particular position and definition in macro-economic planning and development. For most islands agriculture represents a declining economic sector whereby policies are attempting need to arrest it for the purpose of maintaining some level of viability and national contribution to the debt crisis. In islands such as Antigua this argument has all but been lost. In Barbados and St. Vincent plantation and small-scale agriculture still have a recognised role, but the legitimacy of the sector *vis-à-vis* tourism, residential and recreational development, the services, industry and off-shore banking is declining. Nevertheless, it is important to view agriculture and food provision more generally as a significant part of regional macro-economic regulation. Its 'management' in this context allows other forms of development and accumulation to occur. Its marginalisation allows for the selective development of foreign capital, and a growing importation of processed foodstuffs to meet domestic consumption.

Two developments concerning these trends produce increased vulnerability for the Caribbean. One is the pressure for the cessation of preferential access of sugar and banana exports to the European markets (Agra-Europe 1994) not least from retailers and traders in Europe. This gradual process means output and competitive price reductions for producer countries. It can also exacerbate the volatility of balance of payments problems which then can restrict other forms of export growth. In Barbados, for instance, the sugar industry has been in deepening crisis (Drummond and Marsden 1995, 1999). In 1980, sugar accounted for 6.3 per cent of total GNP, by 1992 this was reduced to 1.7 per cent. A second vulnerable factor concerns the outcome associated with the Gatt Uruguay round. The likelihood is that the markets of developing countries will assume greater significance to the export performance of the developed countries. Pressured by the transnational food sector, the latter are likely to insist on reductions on tariff and non-tariff barriers. The gains for the small developing countries are likely to be few. Indeed, while liberalisation is the strongest and most irresistible trend in global trading, its advantages seem to have been captured by those economic interests which can more easily transfer operations to areas where new markets for food are emerging. For instance, the development of Northern retailers and food manufacturers in countries such as Barbados is likely to further constrain local systems of food provisioning; or at the very least, marginalise their sub-economies. In this context, and particularly concerning the food sector the model of development adopted is one which aids the interpenetration of foreign capital such that it provides new markets and opportunities for certain segments of Caribbean society over others. The most powerful interests are strategically shifting capital out of domestic agricultural production into retailing services, food processing, banking and tourism.

It is important to recognise the interlinked nature of some of these trends, and to see how their implications are being played out in different national settings. For many of the entrepreneurial, merchant and private professional groups in Barbados, these trends are seen not only as inevitable but as necessary for the continued development of the island. Further externalisation, built now not so much on the traditional export of sugar, is seen as the best way forward in the context of multinational trade negotiations and the bilateral trade and economic assistance agreements. Moreover, these conditions and logics are reinforced by the conditions of development aid and balance of trade support programmes financed by the multilateral financial institutions. Debt servicing and the need to attract foreign currency earnings reinforce the tendency of seeing agricultural production as if it were like other sectors of production and services. Consider the following quotation, for instance, from the Chairman of Barbados Development Bank:

> *'When we talk about agriculture we talk about sugar agriculture, separating from non-sugar agriculture. What is the optimum acreage should we have in cultivation. The indications would be that the acreages have been on the decline over the years. We used to talk of 60,000 acres, now it is more around 30,000. We would like to raise this to 70,000-80,000 tonnes. But the way things are now, fluctuating. In 1992 we had a bad crop, we came back a little in 1993 with just over 43,000 tonnes. We know for the coming year (January 1995) reaping) it is going to be way down on this year.'*
>
> *'Why I raise all this is the question of the preferential treatment, the agricultural protocol under Lome. What are going to be the terms and the conditions? Again the indications would be that the preferences will be minimised and eventually eliminated. So that's going to throw us out into the open market. So it brings into question the issue of efficiency, and can we produce at sufficiently low cost, efficiently.'*
>
> *'Now as a sort of hiatus you are looking at diversification and there are certain areas coming to mind. Cut flowers is one. There are certain other crops that we can concentrate on, but again we have to look at the perishability of these items, and the transport system to get these to market as quickly as possible.'*

> *'Foreign exchange earnings from agriculture are from all aspects a key factor as far as we are concerned. The way I see it, you know, when you do your economics in the early years you learn about the factors of production-land, labour and capital. But now one of the most crucial factors is foreign exchange. Foreign currency this is a fact.'*

It is worth considering how these so-called structural conditions promote a diversity of local responses. They are far from uniform in themselves, and they demonstrate a key feature of the significance of uneven development in the South. Trade liberalisation and deregulation do not lead to any form of level playing field. Rather, they are responsible for the selective devaluation of certain domestic resources so as to encourage the penetration of external capital and expertise. Endogenous capabilities and resource structures are further 'hollowed out'. In this sense they set the new rules for uneven development and the lack of endogenous co-operation between the islands.

Such processes of externalisation and the exposure of national-based diversification schemes to global market pressures tends to exacerbate the ephemerality of new nationally based agricultural initiatives. Meanwhile imports of food products increase both for consumption by the local population and for the burgeoning tourist and hotel trade. This, in turn increases the pressure by national governments to organise export drives of 'quality' products. As a Barbadian trade commissioner argued:

> *'Our imports are about ten times to one ... Our export figures are not substantial because we are competing with other countries and we have to look at when summer comes in and or winter comes in and then we get our products on and so on ... What I have been doing is trying to get them under contract. So that they (the farmers) are committed by contract to supply at a specific price so they will have consistency of pricing throughout and they are demanded to supply. If you are caught doing those things which are in breach then you can be put out on the limb. The contract should be between the farmer and exporter. Because there is no consistency of supply the exporter has to take it when they get it, so there is a big problem in terms of the consistency of supply. That is where I come in. I believe in order to stay in export we must have consistency of supply.'*

In Barbados there have been a variety of attempts to reorient the agricultural economy around the production of vegetables and fruits, both for the domestic and export markets. This has been a major source of internal conflict and competition between different sectors of agriculture (*e.g.* small farmers and the sugar plantations wishing to replace sugar with other lucrative crops; and older and younger farmers who variably wish to retain a domestic system of provision built upon vegetables staples such as sweet potatoes, yams, cassava, carrot, cabbage). Diversifying the agricultural base in this way has several justifications which are often in contradictory relationships with each other. For the national and international trade and ministry groups it is a strategic opportunity to capture foreign exchange in the context of the debt crises facing sugar. Some of our respondents (small farmers, ministry officials) complained that while it had been exclusively the fault of the plantocracy and the large landowners that this debt crisis had arisen, government and other elites were expecting the small, non sugar sector, to deliver a panacea to this long-running problem.

In the period of study (1994–1995) there were approximately 72 small farmers growing hot peppers mainly for the export market and being encouraged by the Ministry and the marketing organisations. The top quality peppers go to North America and Europe. The rest are exported as sauce or mince, or sold domestically, again according to their quality. As another Ministry official argued in 1994 who is responsible for exports of these goods:

> '*We have increased the acreage in hot pepper. We have increased the export of hot peppers, but because of this increase we found people want more and more in different niches in Canada, Holland, the UK. The requests are more. For all things, yams, okras, hot peppers, sweet potatoes. You find that requests are more per year. So although we increased production it does not go on a par with demand. I am hoping that by the end of next year that we can be up more than 50 per cent on supplying on all these requests.*'

Larger producers become a threat to the opportunities for the use of these markets in enhancing the small farm sector, and in realising some of the state investment placed over the decades in rural development schemes, irrigation and cooperative marketing systems. Many small farmers are sceptical as to the commitment of the state to their maintenance over the medium term. Moreover, the need to provide high quality and regularity of supply and to do so with the minimum of risk means that the larger producers can potentially capture the market advantages. The 'situation of exchange' (Appadurai 1986) in this sense is both highly multidimensional (in that it involves different groups of actors and agencies often for different crops) and competitive (in the sense that the prevailing agricultural conditions are such that small and large farmers are competing for the same export bounty). The networks of production and supply are thus anything but rationally organised or structured and regulated in a coherent way.

While many in government departments and the national agricultural society (which provides extra extension services to small producers) see the need to protect the markets of the small producers and the opportunities vegetables provide for intensive small scale production, the major marketing agencies and trade officials emphasise quality and regularity of supply by the development of much more regulated systems.

particular. In the North, while food demand may have stabilised, the food market is expected to grow at about 2 per cent annually. Improvements are largely through the value-adding process, which often means no more than effectively 'de-industrialising' former 'Fordist' foods (Arce and Marsden 1993). In the South demand is increasing especially in Southeast Asia. As incomes rise for some, pre-packed products of high quality and products which have undergone further processing begin to replace some staple foods, offering retailers significant opportunities. In Southeast Asia the food market is expected to grow by 4-7 per cent per annum. These trends are not just associated with Southern regions but also significantly include the 'new dependent spaces' which are currently outside but contiguous to the regional trading blocks of Europe (former COMECON) and North America (particularly Mexico and The Caribbean). Moving along a trajectory of larger-scale outlets which reduce operating costs and encourage one-stop shopping, retailing world-wide would seem is in the ascendance. Unlike other sectors, however, such as manufacturing, these global trends are not necessarily built upon the global interconnectedness of retail capital. While there is evidence of retail buying alliances and some cross-national activity (e.g. Walmart's entry into Europe), the key to globalised forms of retailing lie in the potential for convergence of national and international regulatory conditions. While these may be principally seen to be the preserve of the national state, retailers through their international and national political and regulatory activities are increasingly influencing the reconstitution of those states. Of particular significance here is the increased national flexibility for retail concentration. The ten largest chains in both the EU and USA account for roughly 30 per cent of total grocery sales; in individual US states and EU member states (most notably, Denmark, The Netherlands, France and the UK) the three largest retail chains account for 40-60 per cent of the grocery market. This not only gives them a strong national base from which to explore transnational expansion (for example, Tesco's announcement to build 20 new stores in Hungary and Sainsbury's incursion into the US) but it also enhances their powers over the food processing and wholesaling sectors, who increasingly have to answer to them on quality and value grounds.

In the British case the rising economic and market significance of the corporate retailers is well recognised (see Marsden and Wrigley 1995; Flynn *et al.* 1994). While global pre-tax margins are relatively low (around 0.5-3 per cent), with gross margins ranging from 20-30 per cent, improvements in efficiency and/or the reductions in costs of goods sold (and/or increases in the turnover of sales) can have a potentially large influence on net margins and profits. Although British retailers have faced pressure recently, some they have maintained and expanded exceptionally high profit margins. These margins are usually interpreted in economically rational ways. These realities, however, are outcomes of a contested process of market maintenance and expansion which has to be deeply embedded into the social and political apparatus of the nation state, and increasingly the EU. As with any type of economic activity, and especially in the case of retailing, given its 'near market' location and traditional reliance upon creating markets under highly competitive and inelastic conditions, it has to rely upon and begin to remould the social and political conditions in which it finds itself.

On the face of it British retailing has been typified by high gross incomes relative to its international competitors through developing efficiencies on sourcing and the rapid development of large stores. This has allowed the efficiencies to be captured 'in house' rather than in passing these directly on to the consumer in the form of lower prices. In addition, looking internationally it can be argued that, for instance, German and North American counterparts have traditionally faced more competition from the discount sector and that British retailers have been more effective at developing distribution technologies.

Probably more significant in terms of these features of the uneven development of international retailing concern the speed at which food commodities *are rotated*. Retailers are not interested in backward or forward integration. Indeed they prefer, increasingly, as we see in the Sao Francisco case, to reallocate the considerable risks in food procurement, transportation and quality maintenance to the other actors and agencies involved. Their prime focus is to increase the speed of turnover as well as to demonstrate product innovation and new value added quality. In the US, for instance, the late 1980s and early 1990s have witnessed a rapid increase in product development. In 1991, 16,143 new products were introduced, with over 77 per cent of these being new types of food. This represented a 100 per cent increase over the previous decade.

Even taking a strictly economic approach this suggests that there is more at stake than margins and outputs. British retailers have given considerable emphasis to reducing the number of days stock are held and in balancing off supplier and banking debts by cash cycles which maximise their control of funds and stock. Here, in parallel to the time-space constraints associated with our production and supply spaces outlined above, it is the speed towards the point of sale which is all important. Efficiencies in the days stock is held has allowed the shortening of the time of internal funds rotation. In Europe, average days stock tends to vary between 25 and 30 days. This has been reduced significantly in the UK and Japan. In addition this factor in uneven retail development works not only internationally, but also intra-sectorally. It increases the efficiency differences between the 'super-league' of retailers and the rest, particularly the ailing independent grocery retailers. In the UK the total number of retail businesses has declined from 142,000 in 1976 to 86,800 in 1990. Further decline ensued in the 1990s in terms of the independent grocery butchery and abattoir sectors. This has largely been at the expense of the specialist independent sector (i.e. the small multiples) and the specialist retailers (butchers, bakers, grocers and dairies). While this trend may be most evident at this point in time in the UK, it is necessary to recognise that it is set in motion elsewhere. In France retail outlets have declined from 200,000 in 1966 to 150,000 in 1990; in 1992, hypermarkets and supermarkets increased shares by 20 per cent and 40 per cent respectively; while they accounted for only 6 per cent of food stores they held a combined market share of 40.5 per cent. Italy, stereotypically perceived as a very fragmented retailing nation, witnessed a doubling of the number of supermarkets between 1985 and 1992, with the number of hypermarkets increasing six-fold. The number of retail outlets has dropped by 15 per cent between 1985 and 1993. These international contextual changes and the largely economic indicators which underpin them raise some interesting questions about the degree of transnational convergence or uneven development in retailing and consumption that regions like Europe are likely to experience over the next decade.

In the British case the continuing dominance of a small group of retailers and the absorption of food markets by them at the cost of independent and specialist retailers is an outcome of social and political factors as much as it is from economic criteria. Moreover, their particular path to success, in relation and sometimes in contrast with counterparts in other parts of the advanced world, has relied upon the particular shaping of the diverse regulatory conditions that surround them. Regulation, in this sense is seen as the process by which power relations (whether within government or beyond) come to be codified and expressed, and through which contested actors and agencies align in order to progress courses of action (see Clark 1992; Hancher and Moran 1989; Marsden 1992; Christopherson 1993). In the British case the forms of regulation which have been significant have developed at least at two levels. These can, in summary, be labelled macro and micro regulation. Both have heavily involved more than simply state agencies, for they have embodied complex networks of public and private actors and agencies. At the macro level we have outlined elsewhere (see Marsden and Wrigley 1995), how the development and potentially the continuity of the neo-conservative state has embodied notions of privatised consumption through the re-privatisation of former state assets (see Saunders and Harris 1990). In this sense, we argue that in the Britain, retailers have at the very least been able to capitalise on the reorganisation of expendable incomes brought forth by the breaking of collective working class interests, and the development of privatised consumer culture at the expense of a growing and in consumer terms, disenfranchised under-class. In addition they have absorbed relatively cheap labour out of the Fordist restructuring of the 1980s which has (not coincidentally) promoted the devaluation of their labour costs relative to foreign counterparts.

In short, the development of the neo-conservative, more liberal, state, both initially in the UK and the US, but now increasingly elsewhere has provided broad macro economic and social conditions which have been conducive to the restructuring of the retail sector along corporate and economies-of-scale lines. Underneath this macro economic and political shift has been the development of particular forms of 're-regulation' (concerning food law, competition policy, private systems of quality control). These have provided the basis for specific nationally based retail capitals to further shape the food supply and provision systems. Food consumption has not been homogenised by this process. Rather, it has become more reliant upon the differential development of nationally based retailers and a redefined state regulatory role. Within the broader raft of deregulatory neo-conservatism therefore, we have witnessed the development of new spatially uneven regulatory domains which begin to foster and promote the corporate retail sectors and the modes of supply and consumption they construct. These domains are far from stable or pre-determined. The combination of food safety scares, devaluation of property portfolios, price wars with the new discounters, as well as the onset of corporatist regulation from Brussels, has tended to threaten the relatively 'benign' regulatory environment retailers experienced for much of the previous two decades. Nevertheless, what is critical is to consider the ways in which retailers, particularly in relation to the nation state, have managed to maintain their market power and, moreover, their public legitimacy in the face of potential moral and political crises associated with the gradual reconnectedness of food, health, environment and consumption concerns.

Since the early 1990s in Britain, following the passage of legislation such as the Food Safety Act (1990) and The Health of the Nation White Paper (1991), which began to shift responsibility for food matters to the retailers, the major retailers have clearly become significant actors in the promotion and implementation of 'public policy' (see Competition Commission 2000). As a result, in a very real sense, they act on behalf of the state in delivering consumer rights and choices. Reciprocally, the regulatory state has in turn become critically dependent upon the continued economic dominance of the retailers in their role as the major providers, *inter alia*, of quality food goods. The state, has shifted from its post-war Keynesian position of regulating structures of provision through macro-corporatist arrangements with producers and manufacturers. Instead, more nuanced micro-corporatist relations, particularly with the service and retail sectors provides opportunities to reorientate and develop food markets and consumption, establishing more *inclusive* rather than *universalistic* provision systems.

Table 2.1 Differential value of foods, labour and environments in dynamic and dependent agrarian spaces

Value construction in food networks	*New agrarian districts*	*Areas of agricultural retreat*
Foods	External quality criteria increasingly rigid and variable; influences upon market entry.	Quality highly variable; dependence on external technologies, knowledge.
Labour	New social carriers. Small farmer niches. Super-exploitation of day labour. New institutional structures enrolling those producers who can achieve consistency in food quality.	Plantocracy still indirectly dominates markets. Small operate secondary markets. Many field management tasks being foregone.
Environments	Increases in productivity influencing growing risks in water-agriculture management.	Water scarcity increases. Land degradation increases. Scarcity of resources constructed.

The last decade has witnessed then a period of retail-regulatory activity which is broadly designed to maintain markets for the retailers and support sourcing strategies based on new free-trade principles providing what seems to be a legitimate mode of consumption for the state and the public interest. In the international context this begins to demonstrate considerable advantages for British retailers over their foreign competitors, and it goes a considerable way in explaining their growing national and international market power. These sets of regulatory conditions also help to shroud the real globalised points of connection and value established between production and consumption spaces by highlighting certain food values parameters (such as authenticity; exoticism and freshness) but downgrading others (*e.g.* real social and environmental production value of products) (see Table 2.1). Retailers are thus important agents both in energising global networks of production and consumption but in also culturally and economically distancing some of the particular social and spatial origins of foods.

Conclusions

The three examples cited here demonstrate different mosaics of regional production and consumption space. They illustrate how the themes of quality, regulation and consumption operate as active and dynamic concepts in creating and maintaining food networks, on the one hand, and different *spatial situations* for those networks on the other. These key themes become influential in creating uneven agrarian spaces which express the inherent contradictory nature of recent agrarian development.

Food increasingly provides a basis for power and vulnerability across and within nation states. In this sense it has enhanced its distinctive contribution to social and political life through its uneven but progressive appropriation of social, natural, as well as, economic value.

The cases outlined here – 'super' quality production in Sao Francisco; regulating vulnerability in the Caribbean; and the power of retailers in the UK – begin to give an indication of how the more regional and deregulated global food sector based upon an agro-industrial model is reconstituting rural space. The agricultural and food accumulation process becomes more distinctive rather than less, with the value of foods increasingly constructed in the post-farm parts of food networks. More comparative empirical work needs to be conducted on examining the role of social actors and institutions at different 'sites' or nodes in the food networks. In this chapter the emphasis has been to begin to question their effects on rural space and how they *create* new patterns of dominant and dependent forms of uneven development. In Chapter 3, we will focus more specifically upon the very construction of the unsustainable conditions with further reference to rural Brazil.

One implication or paradox of the growing reconnections and centrality associated with food and rural space in consumption and quality terms in the North, are the ways this is built upon a growing spatial inequality in terms of Southern production spaces. The revaluation of food in the North does little to encourage local value capture in Southern producer regions (see Table 2.1). Indeed, it tends to increasingly define agriculture as a somewhat detached commodity system, disengaged from its deeper ecological and social base. The processes of globalisation and externalisation now impacting differentially on Southern regions such as these suggests not so much a 'deregulated' system of control and market development as a more variable and asymmetrical set of spatially specific power relations, absorbing some local and regional agricultures over others. Hence, the power relations running through (in a vertical fashion) food supply networks come to dominate more lateral and locally situated networks. To develop strategies for coping with these external pressures it will be necessary for agricultures and food supply to be reintegrated into their local and regional settings.

In analytical terms, therefore, scholars need to begin to see agriculture and food not so much as a highly commodified and commodity based sector *sensu stricto*, contributing to national gross domestic product or relieving debt or balance of payments burdens. Rather, we need to focus upon ways in which agriculture and food are and can be socially embedded in local and regional spaces; as part of local systems and networks. While this is beginning to occur in the North (see Section 3), with a growth of concern to reintegrate agriculture and food with broader aspects of living, as we shall explore in more detail in the next chapter, so far it has been based upon global inequalities in space and value. The paradox of new patterns of agrarian 'globalisation' as identified in north-east Brazil, and the Caribbean, is that it seemingly reproduces highly uneven production and consumption spaces based upon different value parameters. Food travels over long physical and social distances and it continues to remain a highly distinctive economic sector.

3 Global spaces; constructing the unsustainable

Introduction

The globalisation of agriculture and food develops alongside contradictory processes of homogenisation and differentiation; as such it provides grounds for enforcement of regulations and contestations that emerge from the ways in which social actors deal with interpretations and gradations of food, labour and the environment. It is therefore within this political arena that social changes and sustainability of the new agrarian sites have to be accounted for. This chapter, with empirical reference to the new agrarian districts of north-east Brazil, examines in more detail the social transformations occurring regarding changes in markets, production structures and labour practices. The processes underway are severely questioning the sustainability of such developments, and a major dynamic creating and reshaping such production spaces is associated with the asymmetrical value placed upon the externalised quality definitions and parameters of the foods themselves, compared with those associated with the quality of labour or the natural environments in which the foods are grown. Super-intensive production, it can be contended, can no longer rely upon technological fixes - such as those associated with irrigation and water management - to halt the accumulation of both environmental or social risks in the production and supply system. Increasingly, such agrarian developments confront quite profound organisational and regulatory limits which need to be considered if the systems of production are to be sustained. This chapter shows how externalised social value constructions - as a central part of globalisation of food-run through and in rural spaces, of the South, shaping social nature.

In this chapter I wish to explore these assumptions by relating them to the highly volatile and constructed nature of the two major agro-export areas in North-east Brazil (the São Francisco Valley and the Açu Valley). I do this to further our understanding of the interrelationships between three key 'value-spheres' which are currently structuring these 'new agrarian districts'. These are the spheres of labour value, food value and environmental value.[1] As we have discussed in earlier papers (see Marsden 1997; Marsden *et al.* 1996a), these contested constructions in value go some way and to integrate the means by which 'nature' is in a process of constant reconstruction around both the production of 'high value' foods and in their increasingly complex and globalised supply chains or networks.

A focus upon these value constructions, the ways in which key actors shape and define them, and the asymmetry in power they create, helps to demonstrate how these new patterns of globalisation are transforming both social and natural conditions. Contentions and conflicts around these value constructions are dynamic elements in the configuration of such regions. Also, a focus upon these allows a more effective integration between a broader political-economic interpretation of agrarian globalisation (in demonstrating how such regions are created, sustained and become dependent, see Watts 1996) and the (undeniable) significance of social actors operating in local social contexts (see Arce and Marsden 1993). In addition, it also helps to demonstrate that environmental inequalities and uneven development are intricately linked with the social value construction of foods and labour in such *globalised regions.* Indeed, as will be shown, sets of externalised social value constructions 'travel through' the *globalised food networks* in ways which condition, or at least set parameters for, local production relations.

In this chapter, with reference to in-depth empirical evidence from the São Francisco and Açu Valley[2] I show how these different 'value spheres' are continually constructed. In doing this we can demonstrate that global social relations and local social relations (Long 1996) are inextricably *connected through and in space.* That is, vertically through the uneven operation and animation of global food markets, and horizontally through the unequal regional configurations of production and supply.

The question becomes how and in what ways are they connected? In this sense, they are arranged around the different value constructions of labour, foods themselves, and the environmental contexts in which both are materially based. The externalised priorities associated with the promotion of particular definitions of food 'quality' and value which are imposed on such production regions by consuming countries are by no means inevitable. While we show that they provide a central organising and empowering/disempowering framework around which the value and organisation of labour and environment occurs, these asymmetrical relationships are increasingly contested and challenged by other social value constructions. These tensions are explicit in the evidence presented here.

Such an analysis raises some important specific and general questions about the sustainability of the new agrarian districts and agriculture. These regions represent the continuous management of contradictions inherent in the current patterns of agro-industrial globalisation; and they question the continuance over the long term of the current dominance of the consumer oriented food quality parameters in shaping such regional systems. More generally, they inquire whether over the long term, the maintenance of the unsustainable (see Drummond and Marsden 1999) is capable of being regulated.

The following analysis outlines the key ways in which the value sphere of foods, labour and environment are currently being negotiated and assessed over time in the region. It attempts to demonstrate how these different spheres are coming together as key elements in the globalisation of the region. In this sense it provides a commentary on the *social process of agro-industrial globalisation* as it is occurring *in* and *through* agrarian space (see Llambi 1993; Murmis 1993). These *in* and *through* elements need to seen and observed together because they both operate simultaneously, and they are constantly reshaping the social and physical space. Both are influencing the actors, institutions involved proactively and reactively. Thus, from a sustainability perspective both social and physical elements are being reconstituted in an interactive ways. The parameters of labour, food and environmental quality are key mechanisms in this process of reconstitution.

More specifically, in this chapter I put forward these arguments by exploring:

a) the particular social and economic transformation of agriculture and the countryside of the selected region, in an attempt to understand the main features and singularities of that region in the context of globalisation; that is: how nature and food are constructed in that particular milieu.
b) the relevance of this process for understanding the ways in which social categories emerge, how they cope with the dynamics of production and markets; and how they respond to the ongoing pressures of the food markets;
c) finally, the analysis illuminates how sustainable and unsustainable environments and livelihoods emerge in the context of food production and consumption. This is achieved by examining the process of incorporation of these particular sites of production in the food networks, and the recent changes in markets, production, labour and environmental contexts.

by the state to retreat from the sector has proved difficult to achieve. The State has not been able to avoid involvement in the disputes despite the gradual privatisation of supply.

Irrigantes as a social category brought together those who were facing the same problem at that moment, i.e. water billing payments. *Irrigantes* became a suitable label chosen to conceal differences among those producers, in spite of already analysed inequalities and differentiation among them (Cavalcanti 1995). Their moves were also in tune with other social movements, such as the Brazilian rural workers and landless movements. By collaborating with these movements, they benefited from their strategies and people's support, bringing their case to the streets; by camping, protesting, closing the main bridge, going to Brasilia, they make a stronger movement. This forced CODEVASF to allow them to resume the water uses. This led to a cooling of internal problems occurring in the Districts managerial attitude towards the problem.

(iv) Inducing marketised production

A final aspect of the changes occurring in the region concerns the efforts to increase non-seasonal presence of commodities in the export market. The comparative advantages of this region were enhanced by the possibility of several harvests in a year. In recent years, this has been, practically, taken to its highest point:

> '*There will be an increase of 35 per cent in the production of mangoes in the off-season harvest time (48,000 tons in 1997). This is a result of the uses of floral induction, developed by EMBRAPA*' (*Diario de Pernambuco*, 21 May 1997).

What is observed in the field is that, floral induction and other technical practices are very much in use, either to advance or delay production according to external market vagaries.

Transformations going on in the production side is in pace with new approaches to the market; new information technology is increasingly used to attract new foreign buyers for the products of the region. For example:

> '*The Valexport will introduce the Electronic Fruit Auction . This instrument will improve the sales of the products, as well as diminish the costs of transaction via intermediaries*' (*Jornal do Commercio*, 14 May 1998).

This was also stated in interview with the President of VALEXPORT.

The model of development of the San Francisco Valley is in pace with the so called neo-liberal ideas in use, by which it is given prominence to the external market requirements. This also induces both the deregulation of labour and water supplies. Additionally, the highly regulated market for export does not have a symmetrical control on the import of goods. This leaves the national producers without protection from competitive goods in the national market. Furthermore, progress in improving the relations with the market through quality of commodities has been accompanied by changes in the uses and division of labour. Workers are worried about the maintenance of their rights. In the same year of the irrigantes movement there was a general strike of the workers which lasted 3 days. That movement was the first to bring together the majority of the workers around an agenda with which they confronted producers. They wanted better pay, better health and security benefits and the enforcement of rules on the uses of toxics, as related to the workers health.

Conclusion: asymmetrical values and the narrowing of options

Throughout this chapter I have analysed the social dynamics of new agricultural sites by revealing the major ways by which food, labour and nature are being re-constructed. The externalisation of production and the pervading global notion of competitiveness are becoming major parameters for the development of such regions. We have seen also that there are daily confrontations between the major social actors who hold different 'stakes' in the outcomes of such regional development. Gradually the unequal participation of the different actors in the shaping and appropriation of the resources provided, is causing perplexity and discontentment to those who expected to improve their livelihoods. The gaps between social expectations, the physical capacities of the environment, and the prospect of 'sustainable' development are getting wider.

As the last section of the chapter indicates, in terms of market relationships, environmental risks, contestation of water rights and in the significance of job creation, the current sustainability of the new agrarian development looks increasingly suspect. Through our analysis of market developments, changing production structures and labour practices we can see how there has developed an *asymmetry in the priorities of externalised value* associated with particular parameters of food quality. These dominant parameters, moreover, are influencing the ways in which labour and production structures are developing; and in turn how the natural environment is being treated. The faith in technological 'fixes' continues even though there is a narrowing of options available in terms of coping with the balances between food quality, labour and environmental values. Whereas, the early problem to be solved was to increase production through the heavy investment of irrigation systems and planting regimes so as to increase both the area and the intensity of production, this problem, once solved, created others. Organisational innovations were then needed to ensure the consistency of 'quality' production to the export market. This was achieved by the growing dominance of the large agro-export firms and the continued use of agronomic innovations which they could deliver in terms of upgrading the consistency of production over time and in terms of the quality of products.

As we see in this chapter, however, these innovations also reach 'limits' in the sense that they continue to confront both the volatilities in the market - with some firms now striving to regulate not only production, but the growth time of plants even more carefully so as to time the production with specific market demands - and in the labour and environmental spheres. Many of the smaller producers and farm labourers become conscious of these limits in terms of the sustainability of their production systems (seen, for example, in the increasing incidence of pests and diseases and the toxicity risks of labour practices). Water becomes anything but a 'free good', and its regulation becomes crucial for the survival of production. The growing demands of the export market and the lack of protection associated with imports increases the degree of externalisation of such production systems. Meanwhile, production keeps on increasing and the region strives to become a major player in global fruit markets.

As the development proceeds, however, we begin to see, somewhat ironically a gradual narrowing of the options available in solving the new problems that it inherently creates. The development of 'hard' technology is no longer the answer even if it is a necessary prerequisite for continued success in capturing the quality driven markets. Rather, the cycles of production and exchange developed have to rely upon the social construction and organisation of labour and of nature; with rapid and gendered changes in labour practices being matched with organisational innovations in the relationships between small farmers and the export firms. For the latter, the advantages of gaining access to lucrative export markets are tempered with the need to control and police their local production systems. This seems to be becoming much more difficult as both producers and nature begins to 'fight back'.

We see then in tracing these quite rapid developments in the 'new' agricultural regions, the significant difficulties in creating a sustainable natural and social system capable of continuing to generate long-term benefits. The very asymmetries between the values constructions of foods, environments and labour and production processes tends to reinforce this process of unsustainability however progressive and modernising the schemes are projected by development agencies. To study such regions demonstrates the significance of the development of *new cycles of vulnerability* as part and parcel of the new patterns of the globalisation of agriculture and food. Moreover, despite all the justifiable concern surrounding aspects of global climatic changes, in the agricultural and

food sphere in particular, new patterns of regional environmental development (as we see here in North-east Brazil) associated with the globalised and intensive food system are creating just as significant and far more immediate types of environmental and social vulnerability.
The growth in more 'healthy' consumption of fruits and vegetables in the North is based upon the asymmetries; and upon an intensive agricultural paradigm which is conveniently 'distanced' from those sites of consumption. In this sense the globalisation of foods represented by the new agricultural commodities and production systems, are seemingly creating spatially compartmentalised packages or conditions of unsustainability. But at some point, perhaps not to far into the future, new alliances between producers and consumers will signal the end of such asymmetries in the construction of environmental, food and labour qualities. This condition of unsustainability is therefore highly contingent, despite its seemingly inevitable logic.

Notes

1 The concept of value is from hereon used to point out the meanings and gradations that are implicit in the ways social actors construct and explain their world. Labour, food and the environment are thus here analysed as significant and contested value spheres for the understanding of social processes developing in the globalisation of food.

2 The San Francisco Valley and the Açu Valley are two of the most dynamic agricultural regions of the North-east (Araújo 1995, 1997; Andrade 1982); both of them were implemented as part of the development plans and agriculture modernisation policies of the Brazilian government; they rely on the building up of dams on the San Francisco river and Açu and related irrigation schemes. Both of them were transformed by the prospect of export fruit culture. Their development as a pool for fruit culture, however, occurred in a distinct manner. The San Francisco Valley developed bringing together large enterprises *colonos*, who were at the beginning linked to agro-industries; who are being replaced steadily by more entrepreneurial producers who carry on production aimed at reaching competitive markets; although these small farms outnumber the large ones, the most competitive are by far the large firms. The recent development of the Açu Valley is based on large capitalised firms, which benefited 'with the privatisation of the public investment, by which large firms were implemented to use the resources of the river regularisation' (Gomes da Silva 1993, p. 34) because of the failure of the original project for settlement of small farms, small farms have no choice but to associate with the large enterprises. The implementation of a 'flexible' approach to production is a mechanism used by one of the firms in the Açu Valley to integrated small producers under their close supervision on production and commercialisation. Although our main focus is the San Francisco Valley, we bring here a case of the Açu to point out the ways social change and unsustainability are occurring in those new agrarian sites, as their similarities and differences contribute to the development of an approach that allows an understanding of the paths of development in North-east Brazil. both regions are mostly localised in the states of Pernambuco and Bahia, as well as Rio Grande do Norte.

3 The research was supported by the CNPq (the Brazilian National Research Council - which enrolled several research assistants and senior visiting professors from Brazil, Argentina and the United Kingdom - and FACEPE (Fundação de Ciência e Tecnologia do Estado de Pernambuco). Visits to the San Francisco and Açu Valleys also received the support of Fundação CAPES and Federal Universities of Pernambuco, Rio Grande do Norte

and EMBRAPA Petrolina. The authors are grateful to Professor Aldenor Gomes da Silva and Pedro Gama for their support in the field.

4 *Colonos* are those selected farmers chosen according to CODEVASF criterion to receive a modular irrigated plot to cultivate, after the irrigation schemes were implemented in the São Francisco Valley. They were allotted 6 ha with irrigation and other 6 of dry land, *sequeiro*, which could be developed under contract and Bank credit. However, over the years they began to face problems with production and markets and unable to pay their debts. As a result, they decided to give out their plot to others more entrepreneurial and capitalised producers.

5 The work of those graduate and post-graduate students is acknowledged; Juliana V. R. Ramos, Cristina Belo da Silva, Rogerio Medeiros, Flavia Campelo, Morgana Oliveira, Sofia, CNPq grantees.

6 The FRUPEX was a division within the Ministry of Agriculture, located in Brasilia. It dealt with fruit exports. It developed a means of communication and some mechanisms to point out market trends and opportunities. It was closed under the restructuring of the state commenced in the 1990. In 1993, in visit to FRUPEX to Brasilia, we found out it was closing down.

4 Maintaining the unsustainable in the North

Introduction: global trends and rural distinctiveness in the North

An analytical starting point here concerns the recognition of new patterns of diversity and differentiation which are emerging in the rural areas in the UK and in Europe more broadly. This diversity needs to be placed in the context of the general processes which are both undermining and reinforcing local features. The common trends affecting rural areas can be split into those affecting the whole of society and those which are specific to rural localities. Amongst the former are:

(i) the globalisation of many areas of the economy, the exposure of these elements to increased competition, and the related restructuring of enterprises and the State;

(ii) the maintenance and consolidation of a '*liberal market*' ideology which is continuing to strengthen macro-economic management and control in the direction of finance capital and exchange;

(iii) a consequential search and contestation for *new* forms of governance which do not rely upon heavy state intervention and rule by a bureaucratic elite, but rather seek (selectively) more public participation, often mobilised through local and regional (rather than national) systems of governance, partnerships between levels and interests, and greater reliance on private and public 'entrepreneurship' and innovation;

(iv) related deregulation and re-regulation (at different levels and involving different processes (e.g. currently within the dairy sector), and privatisation of formerly State owned or regulated activities;

(v) increasing efforts to internationalise the *terms under which trade takes place,* and removal of economic and other barriers to trade, including most notably production subsidies, for example through the GATT and the WTO;

(vi) the introduction and uneven spread of *new technologies,* especially ICT (information and communications technologies) and GMs;

(vii) *demographic changes,* leading to the ageing of the rural European population and increasing dependency ratios;

(viii) selective but increasing *personal mobility*, including commuting, migration, tourism and recreation;
(ix) changing *cultural values* in relation to individualism and community, responsibility for future generations and the development of democratic practice and citizen involvement;
(x) the emergency of new winners and losers from these change processes, and especially the appearance and recognition of '*excluded groups*' suffering from poverty and various types of deprivation.

These trends are affecting all advanced countries differentially, and they set the context for assessing the policy options available for rural areas. While there is much talk about viewing rural areas as 'holistic entities', which are by their very nature integrative, (arguments portrayed in much of the sustainability discourse) it needs to be recognised that many of the processes of change and development mentioned here are impacting and shaping rural areas and life chances in uneven and contradictory ways. *They are anything but integrative or compatible*. Rather, it will be a test of the success of the new forms of governance (including the Assembly) as to how well it can manage these processes in ways which foster clearly defined *competitive* and *cohesive* objectives for rural areas. These governance structures will have to look *outwards* as well as inwards; they will have to translate international processes in ways which protect the human, natural and physical asset structures of rural areas.

This will be made all the more difficult given the recent trends for rural areas to become progressively *less* self-sufficient and self-contained and ever more open to the wider forces (economic, social and political) shaping European and indeed global development. This progressive lack of self-reliance has not resulted in more uniformity across rural space. Rather, it is promoting more rural differences in terms of the structure and performance between areas. Apparently similar areas can demonstrate quite different characteristics in terms of key indicators like net migration, commuting, deprivation, new enterprise formation, the degree of social cohesion or fragmentation and so on. It is becoming apparent that different rural areas display *distinct capacities* in the face of global markets and social change. This will pressurise the arena of rural governance (in Wales, the National Assembly) to find the ways and means to foster these collection of distinct capacities; to be cognisant of the different productive and consumptive relationships rural regions hold. This will require it to progress the development of an *Integrated Rural Policy* for Wales (IRD) at the same time as it responds to the different regional priorities of different parts of rural Wales. In addition to this it will need to separate some of the immediate problems facing farming and the rural economy from their deeper causes. Tackling the latter will need to be built into a rural strategy from the start.

Before I begin to suggest the possible priorities of policy action for the Assembly, however, it is necessary to specify the particular significance of agriculture and food to the rural economy, as well as to the well-being of Wales as a European region more generally. Policy developments will need to foster the positive contributions of agriculture and food to the rural economy as a significant element in increasing Welsh rural capacities and resources. A *strategy of innovation* is central.

It is crucial in an era when the international economy favours firms and regions which are able to learn, change and adapt faster than their competitors, and where the most appropriate forms of economic, political and social organisation are those which facilitate learning and innovation. The question is how should this be specified and what are the best policies and structures to deliver it. To answer these questions it is necessary to explore some of the new interconnections between agriculture and the wider food industry; and then to pose how both can contribute to a more sustainable rural economy for Wales.

Welsh farming and the rural economy: a region responding to the agro-industrial model

Wales is a predominantly livestock farming region with milk, finished cattle and finished sheep and lambs representing 84 per cent of the total value of agricultural products, followed by poultry (8 per cent) pigs and potatoes (2 per cent each). Agriculture still accounts for 1 in 4 jobs in rural Wales. For the UK as a whole, Wales provides two-thirds of the national specialist beef sector, most of it coming from the Less Favoured Area designated areas (which accounts for 80 per cent of the total agricultural area of Wales (Bristow 1997).

Such statistics do not elucidate, however, the profound role that agriculture and, more broadly, the food supply industry plays in Wales as a crucial element in delivering social and environmental sustainability objectives as well as promoting wider patterns of local economic development. These issues are generally well documented; but they have even more profound significance in the present context of uncertainty and crisis facing the sector. As I will outline, in order to understand and to specify the contours of this crisis, and to persuade people of its depth and seriousness, (farmers are often perceived as 'crying wolf' far too often) we have to view agriculture as part of a complex and changing food supply system. Moreover, as I will argue, to section agriculture off as a separate and distinctive sector for specific treatment is increasingly likely to both obscure and prolong the problems being experienced.

To answer the questions facing the livestock industry in Wales it is necessary to begin to ask some specific questions about the contribution and constraints in viable food supply in Wales; what the potentialities are for harnessing social, economic and environmental benefits, and how the assumed 'marginal' characterisation of much of Welsh agriculture can be redefined and linked to new innovative strategies for rural development. I will divide this summary of the issues into three sub-sections dealing with: (i) the expression of the immediate crisis and context; (ii) the underlying basic causes and trends; (iii) the current pivotal role and position of corporate retailers and their supply chain influences in what I term their new role as 'the masters of the food system'. One of the tendencies in the recent past has been to try to deal with problems in the short term, as they arise, without tackling the longer-term issues. We should remember that the BSE crisis is now well over ten years old and the problems of regulating agriculture through modifying the CAP dates back at least to 1984 when milk quotas were first introduced. In addition, despite some fluctuations gross capital formation in the agricultural sector began to decline form the mid-1980s, representing the end of the post-war rise in capital intensity on farms in the UK. The agricultural sector, most notably in upland Britain, continues to suffer what can only be regarded as a form of 'arrested decline'; whereby the state apparatus continues to project a dependency culture, but one which it finds increasingly difficult to uphold for financial and political reasons.

The immediate conditions

If 1996 reaffirmed the significance of the BSE problem and its deepening effects upon beef markets both at home and abroad, 1997 was the year when a series of other factors combined to reinforce the uncertainty and economic insecurity of livestock agriculture and its markets. These factors include:

(i) growing financial uncertainty and insecurity in the Welsh farm sector with net farm incomes down on average by 23 per cent, more farmers slipping into the red, and prices for finished cattle down by 25 per cent. Increases in the supply of lamb and pig meat only add to the problem, with lamb prices down about 28 per cent and pig meat 15 per cent.

(ii) these conditions, initially stimulated by the BSE crisis, have been compounded by the high level of the Green Pound and the reductions in compensation payments associated with Hill Compensation and Livestock Allowances (HCLA payments in Less Favoured Areas) of the order of 27 per cent since 1992;

(iii) there have been particular uncertainties experienced by specific sectors; for example in the dairy sector the deeper effects of deregulation are increasing the costs and responsibilities of the smaller end of the dairy farming sector and making the continued investment in that sub-sector more uncertain (i.e rise in collection charges and fall in prices of standard quality milk);

(iv) in addition, and an issue which has quite recently come to the fore, has been the growing realisation on the part of the farming community and beyond that imports of beef products have been increasing (not least because of the strength of the Green pound); with prices falling by 12 per cent for French beef, 14 per cent for Irish and Dutch beef products. This has led, further, to the accusations that downstream retailers and processors are able to 'make a profit out of a crisis' with evidence showing in aggregate terms that the 'price spread' (i.e. retail price minus producer price as per cent of retail price) for beef in GB increased from 48.6 per cent in 1996 to 54.4 per cent in November 1997; and that for lamb increased from 36.6 per cent to 51.9 per cent over the same period (Welsh Office Agriculture Department 1998). The price spreads – i.e. the difference between farm gate and retail prices – increased again between 1997 and 2001, with the smaller producers and processors bearing significant regulatory costs associated with hygiene and food safety conditions.

These latter realities have fuelled questions about the 'mark-ups' inherent in the post-farm parts of the food chain, particularly given the lack of evidence of equivalent falls in retail prices, growth in imports, and a perceived lack of 'trust' in food supply chains more generally. Producers are realising that they need to consider direct action towards both the traditional state authorities and, increasingly, the supermarkets and caterers, and their nationally and internationally-based distribution systems.

It is important to appreciate that these combined outcomes are also associated with deeper structural problems which have been set in train over a longer period. Successive attempts to deal with these have largely failed. Added to this should be the recognition that overall subsidies to Welsh agriculture reached £135 million a year (by 1995), representing on average over 50 per cent of total income for a Less Favoured Area (LFA) farm; and the recent attempts to reduce this vulnerable dependence upon the public purse only heightens the degree of acute uncertainty in the farming sector.

Some basic and underlying trends from the past (and for the future?)

Foremost among these is the continuing and profound demise in the effectiveness and legitimacy of the CAP, and its associated supply and production regulation to solve the problems farmers face. Impending World Trade Organisation (WTO) pressures to reduce farmer support, in addition to the reluctance of some European countries to be seen as upholding the existing CAP arrangements, tends to add to a logic of expecting farmers to increasingly fend for themselves (through, for instance, developing diversification enterprises and joining environmental schemes), in the face of the WTO and the annexation of new eastern countries to the European project. This is part of the *'race to the bottom'* scenario which most Welsh farmers will find difficulty in winning. The scenario suggests a growing disparity between the East and West and Lowland and Upland Britain, with those technically advanced farmers being those only capable to compete on world markets while the rest sit waiting for reconstituted rural welfare payments.

In short, the current trajectory of global competitive agriculture is likely to reward those producers who can reduce their costs of production still further. For the majority of Welsh farmers the cupboard from which the resources are taken to achieve this are bare. Moreover, in a context of severe national budgetary restraint concerning public spending it is difficult to envisage extensive positive discrimination for the farming industry towards the smaller end of the spectrum. These combined circumstances suggest that if conditions are left unheeded we will witness the further marginalisation of small family farms, not least from the progressive out-migration of young farm family members to other occupations and lifestyles. By the year 2000, there were 1200 fewer dairy farm holdings in Wales in comparison to 1996. There is some evidence that in general terms fewer farmers sons and daughters are planning to succeed the farm. External circumstances only exacerbate this tendency.

Perhaps less dramatically but just as pertinent has been the increasing salience in food supply chains of the construction and use of food quality criteria as a convention of market entry. The food scares of the late 1980s, the passage of the Food Safety Act in 1990, the rise to power of the corporate retailers, and now the possibility of a Food Standards Agency all point towards a complexity of criteria being placed upon the producer sector concerning not only the physical quality of their products, but the assurance and integrity of their management and delivery to the intensely competitive retail sector. This has been matched with the progressive shift in the creation of value in food supply chains towards the 'end-user.' This means that most of the value can be both added and then redistributed *after* the farm gate, and the gradual superimposition of quality standards only tends to reinforce this trend.

The result of both the profound problems of CAP and the emergence of *privately-regulated quality control in supply chains* (governed particularly by the large retailers) is the realisation that the standards and conventions for market entry on the part of farmers is increasingly highly selective and reliant upon new inter-farm organisational frameworks which are linked into quality systems of supply. Farmers now have even less to gain from 'going it alone'. In many of the intensive farming areas of East Anglia and Humberside for instance, buying and marketing alliances between potato and other horticultural producers have been formed so as to co-ordinate and maintain market entry to the retailer-led supply chains. On their part, the retailers are designing environmentally friendly food supply relationships as a marketing tool, enrolling those farmers who have the capacity and the know-how to deliver consistent products under formal contracts. Of course, not all producers can be involved, but the general trend in the innovation of food supply is for production to be increasingly harnessed to a range of quality standards. This also means that there is unlikely to be any such thing as a complete 'free-market' or level playing field in food goods which farmers can simply exploit. Rather, we see the emergence of a rather slow but decisive shift from a universalistic *CAP regulated supply system* based upon quantitative production targets to a more *privately regulated supply chain* approach which gives precedence to quality control. This places and will place new organisational and management demands upon the farm sector, and it tends to empower the consumer end of the supply chain.

The main point to consider concerning these basic shifts in the regulation of food and agriculture is: How can Wales, as an agricultural and rural region, develop a strategy for leading and innovating in terms of these trends? How can we turn it into a *race to the top* rather than the bottom? The timeliness of these questions is paramount given the severe questions about the social, economic and environmental sustainability of much of Welsh agriculture and the lack of effective processing and retailing infrastructure in Wales which could begin to create triple dividends between maintaining a family farming structure, local value-added economic development and the recognisable production of quality-assured foods to a national and international market.

The obvious immediacy of maintaining a dependency culture on decreasing amounts of CAP support for agricultural production needs to be matched by a strategy for indigenous rural and agricultural development which can demonstrate the positive benefits to consumers, rural visitors, tourists and retailers, that a small farm economy can be sustainable. This requires a considerable switch in the agro-economies of scale arguments which still pervade many of the policy debates.

The pivotal role of the corporate retailers: the emergence of privately regulated supply chains

We need here to consider both the general strategies of the corporate retailers, and more specifically their role in both the maintenance and the potential amelioration of the Welsh livestock crisis.

First, in general terms, we have to recognise the unprecedented levels of growth and rates of return experienced in the corporate retail sector during the 1980s and 1990s. This far outstripped that for manufacturing in general terms and food manufacturing in the 1980s, and was assisted by the abolition of retail price maintenance and the flexibility to operate in the market place without too much regulatory restriction. Other countries, for instance, have put limits on store size and protested their independent grocery sectors.

In the 1990s, however, faced with the food scares, more 'careful consumption' and increased competition from the discount sector, *increased competition* and *innovation* have been the by-words to the corporate retailers pre-eminence. In particular, as the 'quality revolution' has gained pace in the 1990s, they have been particularly adept in projecting themselves as the new custodians of both the consumer and the food-supply chain. It is estimated that they receive 70 per cent of Welsh livestock products, and through their 'due-diligence' obligations, they are able to insist upon arms-length control over the quality of their supply as well as indirectly affecting the destiny of the rest of the products.

While it is now being recognised that this market and regulatory power may not always be in the interests of Welsh agriculture, or the smaller upland farmers more generally, it is important to point out that the innovation and competitive strategies adopted by the retailers in the 1990s have tended to encourage them to place more emphasis upon (i) *intra-competition* with their supermarket rivals through, for instance, product innovation, own-branding and customer loyalty labelling; and (ii) *competing at the local level* to insert themselves further into and down the settlement hierarchy, often by engaging in 'store-wars' battles with their rivals at local level, now in the market towns as well as in the conurbations. One result of this is that in Wales we now have the presence of the corporate retailers in most of the market towns. They are leading the restructuring of food consumption and retailing at the local level.

These two dimensions of retail competition and innovation in the 1990s have meant that the process of securing quality food supply has only at best been one of their key concerns. Their competitive and innovative energies have been placed where the intense competition has been greatest; that is between themselves. Corporate retailers, by dint of the relatively low levels of concentration particularly in the small farm sector and the universalistic approach of CAP support, have not had to seriously compete with each other about which farmers to source from. Neither have they been required to be seen to be acting positively towards particular groups of farmers. Moreover, by putting in place more effective packaging and processing of foods they have indirectly added value down the supply chain, away from the farm gate, and indeed the smaller abattoir. In short, they have established a national and increasingly international *food 'grid'* which extends far beyond their local stores or head offices and incorporates, through selective contracting, local abattoir, wholesalers and processors. They have, until quite recently be intent upon establishing a national distribution system which increases the costs of transport relative to the costs of farm-gate products. When the costs of adding packaging and processing to meat products is also included, one can begin to see how the relative value-added has shifted down the supply chain.

The contrast in the general economic health of the corporate retailers sector vis-à-vis the beleaguered farm sector could not be more stark. The onset of a national food distribution system may have helped those farmers who gain lucrative access to quality supply chains, but for many more it tends to constrain local and regional added-value strategies. Studies on supply chain links in Wales for instance, point to these growing constraints; the inability of Welsh food processors to access sufficient resources to progress market development; the lack of vertical linkages in the supply chain which leads to the widespread view among producers that processors are 'conspiring' to withhold a share of profits derived from processing the output from their farms; the lack of co-ordinated product branding strategies. Moreover, the development of more centralised distribution and processing nodes has reinforced reductions in the throughputs of local abattoir and processing facilities. In the dairy sector there is a shortage of primary processing capacity in Carmarthen and Pembrokeshire and limited resources among smaller companies for market development and establishing cost effective distribution systems for meeting, *inter alia*, the wider retail markets. In the beef and lamb sectors there is a recognised lack of confidence among primary processors; slaughtering 'over-capacity' affecting margins and investment capital; and despite the efforts of 'Welsh Lamb and Beef Promotions' a lack of incentives for macro-improvements in meeting quality requirements of processors and retailers.

Under conditions of general inelasticities of demand for basic foods, the evolution of the retailer-led national '*food-grid*' has tended to reinforce the constraints for the development of regionally based food supply-chain infrastructure. This has made the possibilities for *value-added leakage* from the farm gate and potentially out of the region even greater. Whilst, therefore, in many policy circles the conditions and prospects of the retailing and agricultural sectors have been seen as far removed from each other and separate (with the former associated with a dynamic and innovative service sector, the latter with primary production hooked into a highly regulated and state (CAP) supported system, for instance) it has become increasingly clear as the disparities grow between each that they are critically related through the operation of food supply chain links. And, that these supply chain links hold broader implications for broader aspects of rural development.

Over recent years, as a part of their innovation strategies, corporate retailers have begun to consider local sourcing as a serious possibility in providing consumers with more trust and assurance about their food purchases. In addition, concerning a range of environmental lines involving aspects of Integrated Crop Management (ICM) and various 'welfare' lines, the retailers have begun to source more selectively on the basis of local identity and environmental value. This may be a trend that continues and one which Welsh agriculture can more fully exploit. So far its significance has not been quantified.

It is, however clear from this discussion that Welsh agriculture, and more generally food supply chains that incorporate either Welsh locations and foods, is clearly at a cross-roads in terms of its future development. What are the specific issues regarding retailing and Welsh food which need consideration at this particular juncture?

Specific questions concerning the role of retailers and the Welsh farm crisis

(i) There is considerable confusion about the degree of price mark-up between farm gate prices and retail prices. In aggregated proportional terms the evidence suggests that the retail margins have been getting larger during the period of 1996-1997, and from 1997 to 2007.

(ii) It is nor clear what effects other links in food supply chain make to these price changes? e.g. processing, distribution, packaging plants; or the degree to which exclusive contracting between the various firms effects prices and quality of goods

(iii) Producers are likely to face increasing quality criteria both from government and from the private sector; how can they cope with this and what effect will it have in marketing Welsh products? Some retailers are now suggesting that all farmers will need some form of formal certification or license to operate and sell an assurance to 'the consumer' that they are responsible producers.

(iv) How significant will regional sourcing become in the marketing and innovation strategies of the food retailers? How can a higher level be achieved given the dominance of the 'national food grid'?

(v) How can a *multiple strategy* for Welsh agriculture and food be developed which contains elements of:

  (a) promoting the capacity of the farm sector to engage in contracting directly and indirectly with retailers;

  (b) provides the stimulus for indigenous growth in value added capacity within Wales and the Welsh Borderlands;

  (c) develops more sustainable and organic food supply chains for a variety of local, regional, national and international markets;

  (d) shapes the current levels of CAP and national government support in ways which stimulate the above rather than maintaining a 'dependency and a marginal culture';

(vi) A key issue here is timeliness. It is clear that Welsh agriculture and food cannot wait for the protracted process of CAP reform. CAP reform will be a protracted affair and Wales will have to make effective efforts to press for a reasonable level of modulation in the new pricing structures (including Pillar 2 of the CAP) and to be the recipient of any of the revised structural funds (e.g. the new Objective 2 regions).

The short and medium term aims of the National Assembly: the struggle towards a new agenda

A principal element for the development of Integrated Rural Policy in Wales, and the role of agriculture and food within it, will concern the need to foster combinations of economic, social and environmental benefits. Rather than agriculture being viewed as a distinct and separately regulated sector, it needs to be *re-integrated* into the rural economy and assessed as such. More specifically, it needs not only to be seen to be the producer of quality foods that can increasingly attract a price premium, but also as the *sustainer of a wider range of local rural employment and environmental opportunities*.

This requires a focus upon stimulating economic development along *food supply chains* as opposed to considering agriculture as a land-based occupation in isolation. It follows here that what is required are not simply adjustments to the agricultural support structures but also changes to aspects of broader regional economic and environmental policy. The Assembly will need to consider how the various sectoral policies that impact upon the rural economy (especially agriculture, environment and economic development) can be tempered in ways which deliver and foster more integrated benefits and to provide more positive synergies between the production and marketing of quality foods, the enhancement of the rural environment and the development of indigenous rural employment opportunities in the food economy. Welsh agriculture provides the potential to achieve these goals simultaneously.

This is by no means easy. For instance, there are dangers with the development of the *All Wales Agri-environment Scheme*, which has been built upon the success of *Tir Cymen*, that it will be blind in its operation to the demands of quality food production, the stimulation of organic agriculture, and the types of marketing relationships farmers are participating in. In short, it will need to have wider objectives and contributions than to protect the local farm environment. It should be linked to the multiple objectives of creating *green worth* agriculture and food supply chains in Wales, and linked to the broader definition of sustainable development which centrally incorporates rural social and economic benefits as well as those environmental.

Despite these cautions and complexities in unravelling the inconsistencies and distortions of current rural-based policies, it will be important for the Assembly to take the medium and long-term view. To not be hemmed in, by the more immediate pressures of allocating the existing payments to farmers or dealing with this or that food scare or environmental conflict. This is an opportunity for Wales to take the creative lead rather than become victim to the marginalised 'gridlock' of agricultural politics and policy implementation. It is particularly a time when, through the effective development of an *innovative rural and farm policy*, Wales can develop comparative advantage in social, economic, cultural and environmental terms in a European context. It could, within a decade, turn what has for too long been regarded as a 'marginal agricultural region' (defined as this is according to the old logics of scale economies in agriculture) *into a leading European region of agricultural innovation and food quality. How can the Assembly assist this important medium term goal?*

Food and rural policy in Wales: recent encouraging government reactions

The period 1999/2000 saw the establishment of the Agri-Food strategy for Wales; an innovative attempt to coordinate and develop a more effective government support structure for Welsh food from 'plough to plate' (see Banks and Bristow 1999). Following the preparation of three task group reports – on red meat, dairy and organic sectors – the agro-food partnerships, covering the four sub-regions of Wales were formed, with resources and staff allocated to implementing the strategies adopted for the three sectors.

This approach was quite unique in the UK, it followed a series of recommendations made by earlier government reports (see for instance Welsh Office Agriculture Department 1998), and it provides one potentially good example of how cooperation between the Welsh Development Agency (WDA) and the newly formed Assembly *could* take hold. The first year of the WFS has been to lay the foundations and infrastructure for the effective development of food supply chains in Wales; undertaking audits of existing initiatives and beginning to allocate funding (such as through agricultural marketing grants to innovative value-adding processing in the dairy sector). One of the key aims is to use the strategy for facilitating the drawing down of funding from EU Structural Funding Objective 1; and to fit effectively with the aims of the Agenda 2000 (Pillar 2) rural development regulation.

This gearing up process, however, is being severely hampered by the persistence of the deepening farm crisis that is being experienced particularly in the Welsh farm sector; these include the annual battles between farm unions and government for immediate support and compensation associated with the drop in farm incomes, the recent food and mouth crisis, and the imposition of growing regulatory charges (e.g. food hygiene); and the lack of control that the Welsh governance system has over the operation of the CAP budget. The combination of MAFF control over the 'first pillar' funding (over 80 per cent) of the total EU CAP budget, a farmers union representation which continues to focus upon traditional production solutions to the immediate farm crisis, and an Assembly committee which politically has to deal with this day-to day 'crisis-management' means, increasingly, that the opportunities that could be grasped from such policy developments as the WFS are significantly put at risk.

Moreover, in addition to the WFS there are real problems of coordinating the different strategic planning exercises that are now occurring in parallel. For instance, the Welsh Rural Partnership, established by the former Welsh Office in 1998, has promoted a wide range of integrated and sustainable development goals which have also been espoused in the recently produced Rural Development Plan (NAW 1999a). The latter is being produced in order to satisfy the needs for justifying funding for rural development (the Second Pillar) which has a broader (all-Wales) remit than Objective 1; and incorporates wider rural development issues associated with diversification and community and environmental development. *If the first pillar of the reformed CAP may indeed be seen as holding back the potentiality of the WFS; the 'second pillar' and the associated elements of structures funding (Objective 1, 2 and 3), could potentially lead to uncoordinated and competitive bidding based upon a fragmented approach.*

This is certainly a danger; and it is made more probable by the different spatial boundaries associated with Structures funding and the slightly different time – frames for the submission of the plans and their funding arrangements. This is tellingly intimated in the following passage from the Rural Development Plan (NAW 1999a):

> *'The rural development measures to be applied in Wales have been integrated into a single Rural Development Plan, to the extent that this is practicable given the complex inter-relationships which exist in terms of funding certain measures under the Objective 1 and the proposed Objective 2 programmes, the Objective 3 programme for East Wales, and the LEADER plus and INTEREG Community initiatives. The planning documents required for these programmes and initiatives are at different stages of development; this plan reflects that position, and takes account of the information available, for each of the other documents at the time of submission. The Objective 1 Single programming document and the Objective 3 Operational programme have been received by the Commission and are now subject to negotiation (still ongoing); this plan will reflect the outcome of those negotiations in due course. The Objective 2 SPD and the plans for LEADER plus and INTEREG intiatives are in preparation, but are not due to be submitted until 2000. These plans will be in line with the strategy and priorities set out in this plan'* (p. 9).

We can see here then that in the context of the WFS, and the broader aspects of agro-food development planning and support, *that while there are important synergies to be grasped between these distinct plans and their related funding streams, this demands careful coordination.* It is very much an exercise in the operation of 'multi-level governance' in a context of the persistent crisis of the 'old style' command and control agricultural policy. Multi-level governance places increased pressure upon the

Agriculture and Rural Development Committee of the National Assembly in providing a more coordinative role which interacts *vertically* with Whitehall and Brussels, *laterally*, in terms of its engagement with the other committees of the Assembly (especially the planning and environment committee) and *diagonally* with the sub-regional frameworks which operate in Wales.

So far, however, their main priority has been expressed in responding to the immediate demands of the farmers unions and in attempting (unsuccessfully) to generate compensatory arrangements for specifically Welsh farmers. This has taken up massive amounts of time and energy in the form of negotiations about such initiatives as the 'Calf processing scheme' and 'food hygiene cost exemptions'. Such degrees of subsidiarity around crisis issues in the 'First Pillar' have irritated (to say the least) relations between MAFF and the Assembly, and both of these with the farmers; and in real terms are only trying to alleviate cost escalations on farmers caused by earlier policy failures. In short, elected representatives and their civil servants in Wales have to a large extent become victims of the policy failures in the 'First Pillar', as well as highlighting the significantly different rural and agricultural agendas between Cardiff and London. This is largely a road to nowhere.

Rather ironically, in this sense, the devolution process with respect to agriculture has tended to expose the failures of UK-wide agro-food policies at the same time as constraining the devolved means for dealing with them. Devolution has occurred just at the time when there is increasing recognition that rural regions are becoming more differentiated. That is, the conditions, problems and complexities of rural regions (say between rural Wales and lowland England) are diverging as the regional governance and devolution agenda takes hold. However, the still centric, sectoral and top-down nature of Pillar 1 of CAP (as well as much of its bureaucratic administration) acts as a strong centralising counterweight. This tension, in itself, can act to create an inertia in policy making just at the time when innovative changes are desperately needed. The first year of the Assembly expressed these tensions and constraints quite clearly. *The question now is how can they be overcome such that the new agro-food/rural development agenda can be progressed?*

This gives added weight and significance to the opportunities that structures funding, the Second Pillar, and the revised and more pro-active role of the WDA in agri-food matters can have in developing a new support framework which is largely outwith the current raft of CAP and Whitehall policies. Wales will be able to draw down these funds more independently from Whitehall, even though they will remain at a much lower proportion of total agricultural spend as regards the First Pillar. It is telling that the UK government only expect Second Pillar funding to rise to seven per cent of total CAP funding by 2007, while other member states are pushing for far higher proportions.

First pillar funding in Wales depends upon a more mature set of relationships developing between Cardiff, Whitehall and Brussels. At the moment, for instance, MAFF both acts as a Ministry for the UK on many of the EU/CAP issues, but now also as the Ministry for rural England (see rural development chapters for England). This means that strategic agro-food and rural development policy in Wales needs to be more pro-active as well as more bespoken to the particular sets of Welsh conditions it addresses. This has been the focus of attention of the National Assembly Committee for Rural Affairs in 2000–2001 in undertaking a comprehensive policy review on rural diversification (NAW 2001a). Moreover, NAW needs to be making a stronger European case for the diversion of funds from Pillar One to Pillar Two. MAFF, however cooperative or supportive it may or may not be, may not always defend what might be regarded as a secondary territory.

These evolutionary governance relations will become critical as the European rural and agricultural policies, themselves become more devolved. The 'National envelopes' and 'rural development plans' are only the start of the devolution of the CAP and its widening to adopt rural sustainable development principles along subsidiarity lines. This places more emphasis upon the need for careful strategic and integrated thinking regarding rural policy in European regions such as Wales as well as more assertiveness on the part of the National Assembly concerning engagement with the UK ministry and its role in negotiating British rural interests in Brussels.

Conclusion. From crisis to joining-up policy: food, agriculture and the rural contribution to a sustainable Wales

Despite these significant constraints (and opportunities), the adoption and diffusion of a more effective 'supply-chain' regulatory and support system in Wales is being gradually diffused into the new plans and policy initiatives. However, this is tending to only exacerbate the policy differences between these and the farmers unions. The test of the WFS will be in its developing coordinating function, and the degree to which *it can dominate the policy priorities and agenda associated with agriculture and agro-food developments*. Moreover, it will need to connect with the broader, non-agricultural rural development lobby (most notably articulated in the Welsh Rural Partnership); which often holds a decidedly anti-agricultural stance given the recognised historic disparities between levels of public support, its real (negative) rural development impact, yet the growing need for more effective non-agricultural rural welfare support.

A major task will be in making sure the broader rural development policy objectives are met, and that agro-food is seen to play a new and key role in helping to deliver these. In this sense the debates need to move partly away from 'the farm', and towards rural businesses (including those associated with food). Stimulating rural entrepreneurship will need to be a central goal of all the policies and initiatives. Evidence from assessing rural development initiatives across Europe suggests that the farm as a multi-functional business is emerging in many rural regions. This provides an alternative to the economies of scale approaches which tend to view the disappearance of farms who cannot compete in scale terms as inevitable. One key element of this transition is, however, the existence of lasting support networks with regional institutions including government.

The political and strategic leadership of the National Assembly will be crucial in grasping the potential of these developments. Moreover, the environmental planning system and the need to adopt sustainable development principles (an obligation of the National Assembly) is not an unrelated area of policy development. Will, for instance, national environmental planning guidance dovetail with the differing spatial zoning now occurring in relation to different Objective funding, or the WFS and WDA sub-regions? How will the final version of the spending plans for Objective One link in with the European Spatial Planning Perspective espoused by Brussels? Will National planning guidance stimulate or curtail new rural development initiatives spawned by Objective funding and the 'Second Pillar' (i.e. the rural development regulation funding)?

In short, despite a growing consensus of intent, there are always real dangers that these different arms of rural development policy will lack policy synergy. The gradual demise of the old sectoral approach to agriculture is giving way to a new policy terrain. This is much more explicitly built upon a philosophy of integration and multi-level governance. But we can see that fragmentation and a lack of central coordination are real dangers, however 'joined up' the intentions of each policy package is intended to be.

These are some of the new 'joined-up' questions that will need to be asked about current rural policy processes. Relating to food supply chains, a key process here concerns how a more effective *food supply chain approach* to agro-food policy can become implemented across the uneven rural space of Wales. This involves coordination and integration of objectives between the various actors, agencies and initiatives; and, a recognition of the most effective ways of implementing policies according to the different spatial levels. The policy processes will also need to confront the continued agro-industrial pressures which emerge from the onset of GM and the maintenance of the retailer-led food grid, outlined in the first part of this chapter. New struggles and coalitions between policy-makers, social and natural scientists, and food-chain actors are thus an implicit element in defining an alternative rural development framework for such regions as Wales. The very social mechanics of these frameworks are addressed in Section 3 of this book.

Preface to Section 2
The emerging post-productivist dynamic

Having explored the ways in which the agro-industrial model is constantly reshaping some of the agrarian spaces in both the south and the north, this next section focuses on the emergence of the 'post-productivist' countryside in the advanced world. As discussed in Chapter 1, this holds its own distinct logic. Like the agro-industrial model, it is one which both marginalises and centralises rural places and peoples in different ways, and according to different criteria and rules. It particularly tends to marginalise the agricultural productive sphere, as well as broader food supply chains, due to their decreasing economic significance of the sector to rural society on the one hand, and the increasing pathological nature of intensive production on the other. In short, the farm becomes a criminalised space, a place where the 'dirty business' of intensive agriculture occurs usually behind closed doors and gates. Alternatively what becomes positive in the post-productivist paradigm is public external access to the farm resource, particularly the aesthetic character of the agricultural landscape, and the regulation and restrictive planning of this for the broader symbolic public good. The emphasis is on the growing consumption orientation of the countryside (see Marsden *et al.* 1993; Murdoch and Marsden 1995), and the regulation of the 'surface' of this countryside for specifically selected social groups. It very much treats the countryside as at the very edge of the 'suburban frontier'; with the welfare and degree of sustainability of the countryside increasingly tied to the outside, urban world of both work and play.

The power of this model is now witnessed in many of the social-science and key policy-making agencies associated with governing rural space. Here two points need to be made, by way of introduction. First, the post-productivist model tends to represent, as does the agro-industrial model, a particularly socially constructed view of nature. This involves selective and exclusive concepts of private or restricted and controlled uses of rural space as well as an emphasis upon the aesthetic. Second, the model, again as with the agro-industrial variant, tends to be reinforced by an intellectual discourse which has become increasingly dominant in governance terms. In this case, it is a particular discourse of social economy, social capital and environmental sustainability. These concepts are outlined and examined in detail in Chapter 5. Moreover, if the passwords of the agro-industrial model are ones of volume, standardisation and technology; then those for the post-productivist countryside are associated with differentiation, experience and social organisation.

The following three chapters seek to portray examples of the social science of post-productivism. First, Chapter 5 explores contemporary theorisations of the British countryside from the post-productionist perspective. These theorisations attempt to understand the more recent processes of social and economic differentiation that develop as a result of both the redefinition of agriculture and the uneven processes of both demographic and economic dispersal and decentralisation of the urban economy.

In Chapter 6, we examine these frameworks by returning to one regional countryside: rural Wales. This tracks the evolution of the 'clientelistic' countryside; from its strong agriculturally productivist post-war base to a more tentative regional state based upon agro-food networks and rural development.

In the final chapter of this section (Chapter 7), these issues are explored in more depth in relation to their effects, more specifically, upon agriculture. In addressing the question *How is agriculture coping with the post-productionist consumption countryside?*, we begin to see the ways in which it is becoming, again unevenly, re-integrated into the socio-economic fabric of a more consumer-oriented culture.

The section attempts, therefore, to apply post-productivist perspectives to rural social theory (Chapter 5) and then apply these theoretical precepts to a particular type of (clientelistic) rural region (Chapter 6) and then to the agricultural sector (Chapter 7). What emerges is a richer analysis of the differentiated, post-productionist countryside which counters the more political-economic analysis of agriculture and the state witnessed in the agro-industrial model outlined in Section 1. From the post-productionist perspective, the consumer has been let into the theoretical 'black box' of rural change. The next three chapters begin to explore, therefore, ways of opening up this more complicated, and consumer-led, rural black box.

5 The post-productivist social economy of rural space

'*It is unthinkable that this pleasant land should be allowed to be irreparably defaced . . . rapid progress in recent years of urbanisation, the natural ambition of the town worker to have a house in the country or at the sea-side, the break up of the large estates for building development, the extension of traffic facilities . . . and the eruption in places of beauty, of ill designed houses and shanties all constitute a real menace to the preservation of its natural beauty*' (1929 National Park Committee in Countryside Commission 1995)

'*We have pointed out... that we regard the countryside as the heritage of the whole nation and that it cannot be 'preserved' unless steps are taken to afford those who live in the country adequate opportunities to fulfil their duties as citizens. We point out the not infrequent clash between urban and rural mentalities... we now deal with the conditions which must govern the enjoyment of the countryside, especially by town-dwellers*' (Ministry of Works and Planning 1942)

The rural consumption experience

The perceived threats of increasing visits to the British countryside are, as these two quotations show, long-standing. Indeed, the twentieth century in Britain was partly characterised as a period when, through counter-urbanisation and more regular visiting, the countryside has been increasingly seen as a multifaceted consumption space. A place to recharge the batteries amid the intensity and uncertainty of expanding urban life and culture.

While agriculture was both economically and politically dominant in the countryside, however, that is up until the mid-1980s, the use of rural space for amenity and leisure was rendered only of partial concern. Once the productivist policies and frameworks surrounding agriculture started to fade, however, the field gates have been increasingly, if reluctantly, left open for the possibilities for further access and consumption of the rural experience. These opportunities have been matched by a growth in the diversity and intensity of demand for amenity. Rural localities are increasingly being seen as opportunity spaces for the mass population. This chapter will argue that as a result of structural shifts in the rural economy more broadly, new sets of consumer relations are developing and being expressed in the countryside. This is having the effect of reshaping the function and centrality of rural space. It is contributing to the continuing social construction of rural space. While the state has, since the 1940s attempted to create designated areas of different sorts to soak up these growing demands (such as National parks, Country Parks, Community Forests and the like) it is now clear that informal types of visiting and recreating are features of the deeper countryside. Areas which were traditionally for rural folk.

Sixty per cent of British people visit the countryside on day visits at least once a year. Some forty per cent of the population are regular visitors, going at least once a fortnight. There were 590 million visits to the countryside and 90 million to the coast between April and October 1993, equivalent to one billion visits per year (Countryside Commission 1995). Most trips are close to home, with over half being made by car and forty per cent taking place at the weekend (some nine million visits per average weekend). The most popular activities are walking, eating and drinking out, taking part in sports, visiting friends of relatives, or visiting a leisure attraction or place of interest. In addition, some 76 million holiday nights are spent in the countryside, villages or small towns. Nearly half of the 17 million overseas visitors to Britain in 1989 walked in the countryside. Two thirds of overseas visitors mention the beautiful countryside as a reason for their visit.

It is estimated that some 75,000 hectares, or 6.6 per cent of the land area of the countryside is available for recreation by legal right or some other clear mechanism, for instance as country parks, common land with access rights, Forestry commission, National Trust, Water Authority and National Park land open to access. This is being broadened considerably through the new legislation which will increase access particularly in upland areas. In addition, there is linear access provided by 200,000 km of public rights of way in England. Although the number of visitors has grown relatively little since the late seventies, the amount spent by the tourist and the range of activities pursued has increased significantly. Providing leisure services in the countryside has created over 500,000 jobs and over 15 per cent of farmers are engaged in the provision of tourist enterprises, accommodation and holidays. This rises to 23 per cent for the West of England and 35 per cent in Cornwall. The growth in the range of opportunities to take part in sport has been striking over the past two decades, such that 12 per cent of visitors to the countryside take part in a sport. There is an expanding spectrum of novel activities associated with informal recreation and commercial facilities. These range from the development of theme parks, holiday villages, golfing complexes, holiday cabins, and war games. There are now 400 sites for the latter activity, with rural landowners earning up to 15000 pounds per year for a good facility. These trends are set to continue, being stimulated both by the increasing demand for leisure and consumption of rural resources and the continued decline of traditional rural economic activities, particularly agriculture. Rural tourism and leisure are being increasingly seen by government and rural development agencies, local planners and promotional groups as a main panacea to the problems of rural deprivation and unemployment. And the recent crisis of 'foot and mouth' in the upland livestock sector which forced the closure of rural walkways and tourist spots demonstrated the overwhelming financial dependence rural areas now have on tourism as opposed to agriculture. Most of the peripheral rural regions are actively attempting to attract the tourist and the visitor, and to develop facilities which will encourage such. The attempt to attract more (and particularly richer) tourists is becoming an important arm of local rural development strategies and of national rural strategy. Tourist and recreational facilities represented substantial proportions of investment in the structural funding measures for rural areas and in other local authority – based incentives. In the Midlands Peak District and in Cumbria local initiatives are attempting to develop employment opportunities principally around the forecasted increase in tourist demand. A common view held, for instance argues:

> '*The long term prospects for agriculture seem fairly poor. Much will depend on the success of diversification initiatives but it is difficult to believe that these can do more than slow the rate of decline in land-based employment. The prospect for tourism seem better but much will depend upon the high quality of the product*' (Cumbria County Council 1997, p. 16).

In Wales, the continued decline and uncertainty concerning the agricultural base, particularly following the BSE crisis, is meaning that stronger attempts are developing to attract tourists from home and abroad. The Wales Tourist Boards' 2000 strategy, shows that tourism already represents a significant contribution to the rural economy and can help farm diversification. Domestic UK residents took an average 9.4 million tourist trips to Wales during 1995. These lasted an average 39.6 million nights, with spending increased to over 1 billion pounds in 1995 and with most growth being recorded for holiday or business spending. In contrast, agricultures' contribution to the Welsh economy represented 4.7 million and the food and drink sector as a whole 1.2 billion. The Welsh Tourist Board has developed a Welsh Country Holidays programme aimed at providing 'a range of attractive rural tourism products to service the overseas, short breaks and activity tourist markets'. In collaboration with LEADER groups they are looking to promote the 'quality product – Welsh Country Holidays' and to attract the foreign visitor. In 1994,

690,000 overseas visitors spent five million nights in Wales with an estimated expenditure of 190 million pounds. The integration of the consumption of Welsh foods is also being stimulated by the Leader programmes and the adoption of a Welsh Food strategy by the Welsh Office.

In regions like Wales, with a traditionally marginal agriculture and fragile rural economy, greater emphasis is being placed upon 'selling the environment of Wales', increasingly seen as its greatest asset. This means that tourism and recreation become a major element in rural economic development, and it is predicated upon the continued opening up of rural space to external groups and a wide range of marketised commodities and services. Tourism related employment in rural Wales rose from 7.7 per cent to 11.1 per cent, between 1981 and 1991, employing over 27,000 people (excluding the self employed). The Welsh Tourist Board has provided £32.6 million in assistance since 1985, creating 4,300 jobs.

Increasingly, and particularly in the rural regions which are continuing to experience insecure local economic structures and labour markets, we witness the attempts by the whole array of agencies and public bodies to redefine the identities of rural areas in ways which will enhance their attractiveness to the consumer and the tourist. This process of marketisation and commodification has been linked by some commentators to the growth of post-modern lifestyles and image-building (see Urry 1995), and to the development of consumer culture as a specific contemporary form of material culture in the late twentieth century (Lury 1996). And there has been a growth in the academic literature which emphasises the significance of post-modern lifestyles to the growing variability of these different consumption experiences. For instance, Harrison comments that:

> '*Theme parks, medieval Faeroes and feasts, pop festivals and living museums provide the opportunity to temporarily adopt identities which have new meanings for their participants. As part of post-modernism this dismemberment of group norms allow people to lead eclectic lives unshackled by the legacy of tradition or collective expectation and to respond freely to the market place. But in turn, these consumers are fickle in their loyalties and demand is less easy to satisfy or control*' (Harrison 1991, p. 159).

These statements are now common in the social science literature (see Urry 1995, 2000) and they suggest the shift in cultural values and the growing centrality of the rural experience in them. Emphasis is placed upon the different types of 'sociations' experienced, for instance by the different segments or fractions of the service classes, and the specific types of experience sought.

In contrast, I wish to consider a different perspective associated with the effects of these consumption experiences upon the differentiated nature of rural restructuring. In short, what are some of the effects from a rural restructuring standpoint? What contribution is the non-residential outflow, interest and activity of people to the countryside making to the changing character of the post-productivist and differentiated countryside? This is a salient question, at least from a British perspective in that the onset of tourist and leisure demands upon the countryside have played very much a secondary role in the analysis of rural differentiation in the post-productivist period of rural change (i.e. post 1985). Despite organisations like the Countryside Agency and the Council for the Preservation of Rural England (CPRE) recognising the increasing tensions and conflicts associated with the influx of tourists to the tranquillity and 'quiet enjoyment' of the traditional countryside and its pursuits, there has been little attempt to consider how these shifts are making rural places *more different from each other* or changing the economic and social character of rural places. One suggestion of a positive kind is that it should be contributing to more social and economic reconnections between visitors' largely urban existence and the significance of natural and rural worlds in their daily lives. One could assume that in general terms the trend towards more rural experience should be positive, at least for the visitor if not for the rural resident or landowner.

Regional ruralities as consumption spaces

Recent work on studying rural restructuring and the increasing differentiation of rural place, suggests that rural areas are becoming more regionally situated and dependent upon different types and combinations of production and consumption relations (Murdoch *et al.* 2002). And, while there are national and indeed international trends which suggest the development of the consumption countryside, its shape and effects vary considerably concerning the particular social and regional configuration of that rural space. Also, and this is distinctively associated with the nature and attraction of consumption, it will also vary according to the attempts by rural people and institutions (such as National Park Authorities, development agencies and Tourist Boards) in particular rural spaces to articulate and represent conceptions of nature and rurality to their potential urban audiences; to enrol consumers to particular types of activity and to particular types of socially constructed nature. We currently witness this by many local agencies attempting to re-identify their rural spaces with reference to combinations of natural and social authenticity. 'Cookson country' in Northumberland or, more traditionally Beatrix Potter country in the Lake District. Natural, rural and other social constructions are increasingly used to label place and to market place on the basis of its cultural past/present. Old industrial sites (such as early steam-powered pumping stations) are particularly popular, and many of these are encouraged through the recent possibilities of lottery funding for heritage and recreational projects. These general re-culturation trends restructuring. Commoditisation represents a variety of social and political processes by which commodity values are constructed and attributed to – in this case – rural and agricultural objects, artefacts and people. It does not represent one all-encompassing process which, for instance, transforms agricultural labour processes or simply puts a price tag on a formerly publicly – owned piece of property. Rather, it is important to see it as a diversely constructed phenomenon around which rural development processes coalesce and then diffuse. The question it poses in the complexity and diversity of contemporary rural space is ' how do commodity and other social values shape social practices?' and 'how are commodity relations and values generated, and challenged, through the active strategising, network building and knowledge construction of particular producers, consumers and other relevant users of rural resources?' It involves exploring how people, objects, and combinations of natural and human artefacts are actively assembled in rural places, and then how these are used and valued in market exchange. Conversely, it is also important to understand how some actions remain outside the commodity form. Many local action groups in rural areas assemble resources in order to fight these 'commoditisation battles' with developers and planners. Murdoch and Marsden (1994) have documented at length how in the *preserved* countryside of the South East these middle-class groups have held a strong influence in shaping the social and physical spaces in which they reside.

Below this generalised level of discussion concerning commoditisation, however, it is necessary to examine in more specific detail how social actors assemble and construct 'value' in rural space, and then assess the degree to which this seems to take on a commodity form. Explicit struggles may occur over the attribution of social meanings and values, access to resources and in relation to issues and differences of social identity. Hence to explore commoditisation processes is to examine the means and strategies by which social resistance occurs around prevailing commodity definitions and practices. As with the discussion of regulation above it is by no means a predictable process. Rather, it highlights the socially active nature of what, at a distance, might be seen as economic relations. Hence, it is a central element in the changing social economy of rural space.

In some long-standing empirical work on rural areas in the UK (see Marsden *et al.* 1993; Murdoch and Marsden 1994; Marsden *et al.* 1996b) the multidimensionality of this process for different rural spaces has begun to be documented. This provides a useful comparative device in understanding rural differentiation. The commoditisation of redundant farm buildings, for example (see Kneale *et al.* 1991), was highly spatially variable in the late 1980s. Similarly, the development of recreational sites, such as golf courses, tended to cluster in the *preserved* countryside of restructuring standpoint? What contribution is the non-residential outflow, interest and activity of people to the countryside making to the changing character of the post-productivist and differentiated countryside? This is a salient question, at least from a British perspective in that the onset of tourist and leisure demands upon the countryside have played very much a secondary role in the analysis of rural differentiation in the post-productivist period of rural change (i.e. post 1985). Despite organisations like the Countryside Agency and the Council for the Preservation of Rural England (CPRE) recognising the increasing tensions and conflicts associated with the influx of tourists to the tranquillity and 'quiet enjoyment' of the traditional countryside and its pursuits, there has been little attempt to consider how these shifts are making rural places *more different from each other* or changing the economic and social character of rural places. One suggestion of a positive kind is that it should be contributing to more social and economic reconnections between visitors' largely urban existence and the significance of natural and rural worlds in their daily lives. One could assume that in general terms the trend towards more rural experience should be positive, at least for the visitor if not for the rural resident or landowner.

Regional ruralities as consumption spaces

Recent work on studying rural restructuring and the increasing differentiation of rural place, suggests that rural areas are becoming more regionally situated and dependent upon different types and combinations of production and consumption relations (Murdoch *et al.* 2002). And, while there are national and indeed international trends which suggest the development of the consumption countryside, its shape and effects vary considerably concerning the particular social and regional configuration of that rural space. Also, and this is distinctively associated with the nature and attraction of consumption, it will also vary according to the attempts by rural people and institutions (such as National Park Authorities, development agencies and Tourist Boards) in particular rural spaces to articulate and represent conceptions of nature and rurality to their potential urban audiences; to enrol consumers to particular types of activity and to particular types of socially constructed nature. We currently witness this by many local agencies attempting to re-identify their rural spaces with reference to combinations of natural and social authenticity. 'Cookson country' in Northumberland or, more traditionally Beatrix Potter country in the Lake District. Natural, rural and other social constructions are increasingly used to label place and to market place on the basis of its cultural past/present. Old industrial sites (such as early steam-powered pumping stations) are particularly popular, and many of these are encouraged through the recent possibilities of lottery funding for heritage and recreational projects. These general re-culturation trends mask significant variations in rural consumer culture and its impacts on differentiated rural restructuring. We can indicate this by reference to at least two significantly different regional rural types.

During the past two decades, marked by the growing economic, environmental and political crisis surrounding agriculture (and more recently food), the role of theory for the rural economy has both shifted and been reconstructed by different groups of scholars less tied to, and sometimes highly critical of, the earlier agricultural productionist paradigm (Marsden *et al.* 1986). This chapter will focus on some of the key theoretical advances from a broad post-productionist perspective during this more recent period; showing the utility of theory for guiding research, policy and more effective interpretation of the contemporary rural economy. This is a period when new theoretical advances here seem to be needed, and moreover, when human geographers learn to live with theory much more comfortably. Such developments, however, as in the post-war period, tend to reflect as well as guide the real changes that have occurred in British rural space recently. In particular, the rural 'merry-go-round' with agriculture at its hub looks decidedly shaky.

Moreover, rural geographers can no longer so neatly divide the chapters of their textbooks along the main land using activities of the rural economy (e.g. agriculture, forestry, housing, conservation). The processes of economic restructuring which have taken hold in rural space over the past two decades do not necessarily abide by these categories of development, neither can they be effectively regulated by the national post-war policy compromise which prioritised agricultural production and urban containment. The period of progressive productivism has given way to what has been termed a 'post-productivist countryside'. This requires new theoretical understanding. The rest of this chapter will outline the main dimensions associated with recent rural economic change.

Theories for the post-productivist countryside

Since the early 1980s it has become increasingly clear that rural areas, both in the UK and in the advanced world, have been caught up in a much more complicated national and international political economy: a period of social and economic restructuring which has become highly diverse and fragmented. This has not meant that everything before it has been gradually swept away. For instance, as we saw in Section 1 of this book, the growth in the intensity of agricultural production, of technological advances such as genetic manipulation, and in the drift of agricultural labour from the land, have all continued in many parts of Britain and mainland Europe. Nevertheless, the central organising frameworks established in post-war times, themselves based upon national strategies for urban as well as rural living (Hall *et al.* 1973), have been largely overtaken by the tide of a rural (and urban) restructuring process which has been both economically and socially driven. Before looking at theoretical approaches to these processes it is worth identifying some of the main features from a British perspective.

This can be done in general terms by approaching rural restructuring from both horizontal and vertical dimensions. Probably in the starkest of fashions, at the horizontal level there has been a radical urban-to-rural shift and dispersal in both population and economic activity. These have worked together, with waves of ex-urban groups, such as commuters, retirees and tourists unevenly creating demands for wider and more diffuse forms of rural service activity.

In addition, the relocation or new formation of industry and services has in some areas taken the lead in attracting population growth and counter-urbanisation. The conventional forms of territorial planning have been unable to 'contain' these processes, as they have spilled over green belts, the established new towns, and socially and economically reconstituted many rural villages. This has redefined the relationships between town and country, placing rural space very much in its differential regional context; that is as an essential part of regional territorial change.

Second, and this has been given considerable attention by scholars recently, the vast majority of rural land – because of its agricultural occupancy – is still hooked into a series of distant agricultural and food markets dominated increasingly by powerful and concentrated manufacturing and retailing sectors (Ward 1990; Flynn and Marsden 1992). In particular, the role of the 'downstream' sectors in shaping – at a distance – the rural land-based farm sector is a significant feature of recent agricultural changes. The current raft of policy measures emanating from Brussels and Whitehall, in terms of both production and environmental support, still focuses upon the land-based farmer as the principal recipient of public funding and the delivery of public goods in the countryside. Yet, it is increasingly the non-agricultural actors which need attention in assessing land-based rural change and the vibrancy of the rural economy more generally.

The starting point for theorising about contemporary rural restructuring is to see rural change as the *spatially variable condition* which combines these vertical and horizontal dimensions. Indeed, much of the work over the past decade has focused upon how these processes are being organised in different localities; both by the combinations of actors *in situ* and by their sets of relationships with non-local actors and agencies. Such a perspective – broadly defined as new political economy – has begun to redefine the concept of rural space and rural restructuring itself. Rural space becomes a highly elastic phenomenon, constructed out of combinations and layers of social, political and economic relations, traversing different physical spaces at any one time (Mormont 1990). It is differentially tied to the regional and international economy as much as it is to more traditional forms of national regulation (see Lowe *et al.* 1993).

To begin to theorise about contemporary rural change requires, therefore, a consideration of differential (consumer-led) uneven development. This is only partly defined by the vagaries of agricultural change. It is just as associated with development processes which *combine* former economic sectors (e.g. housing, minerals, agriculture, amenity) in new and often novel ways. One example would be the development of farm conversion markets and more complex property uses associated with access agreements and farm tenure arrangements. During the period of agricultural/industrial modernisation, rural areas were, to all intents and purposes, regarded by the majority of the urbanised population of countries like Britain, as marginal places; backwaters to progressive industrialism. Now, after at least 15 years of uncertainty in the policy arena, and 'boom and bust' economic cycles, perhaps somewhat ironically, some rural areas are seen as much more central to people's lives. Food, the environment and the pressure for amenity are creating new and uneven demands on rural space, such that it is to the accommodation of much broader consumption concerns, rather than just to those dealing with the enterprise of production, which start to foster new types of rural and regional development. These new demands, for 'quality' food production, public amenity space, positional residential property, areas of environmental protection, and for the experience of different types of rural idyll or *urban anti-thesis*, are now much more entrenched in rural space than a decade and a half ago. Moreover, they owe their origins to a disparate array of forces both within and beyond state control or guidance. This means that our theoretical approaches have to be adaptive to a more complex and realigned understanding of state-society relations; a situation whereby, agricultural and planning policy now only provide part of the justification for development in the countryside.

A key theoretical question therefore must be: what are the different spatial expressions of this new rural differentiation, and are there common factors which are actively making rural space different?

First, a typology detailing this differentiation will be outlined. In the second part, the chapter will examine some of the common dynamics of this differential rural change. It should be stressed here, that while the reference point is the UK, there is emerging evidence of similar features developing both in parts of Northern Europe and North America (see Marsden 1995; Marsden *et al.* 1996b). The theoretical developments outlined in the rest of this chapter thus raise important questions for international comparison as well as for differential rural change in the UK.

Differentiating rural change: emerging ideal types

Through theoretical development and ongoing empirical work it is necessary for analytical purposes to consider the nature of rural differentiation through the development and refinement of typologies which attempt to capture the dynamic nature of rural change. One such typology is developed here (based on ideal types for the purposes of situating the rest of the discussion, and in demonstrating the process of differentiation as an economically and socially active force in rural change.

A major theoretical and empirical challenge now faces scholars concerning the need to provide analyses of the rural uneven development. One way to start this is to consider emerging *ideal types* – what Weber (1949) termed *Gedankenbild* – which characterise the range of outcomes expected to the key economic, social and political processes shaping the countryside. Such ideal types have to incorporate combinations of local, regional and distant relationships; and consider how these become established in space. In addition, they are associated with the social and economic *reaction to* as well as the *articulation* of economic change.

Four ideal types characterise the British countryside: the first, the *preserved countryside*, is perhaps most evident in the English lowlands, as well as in the attractive and increasingly accessible uplands. These areas are characterised by established preservationist and anti-development interests and local decision-making. While agriculture has recently been intensively practised, farmers are now seeing the benefits of diversification and in serving the local demand from ex-urban groups. Concerns are most clearly expressed through middle-class fractions (Cloke and Thrift 1990), who may impose their views through the planning system on would-be developers (Abram *et al.* 1996). In addition, demand from these fractions provides the basis for new development activities (often of a very exclusive kind), associated with leisure, industry and residential property. Rural change is thus a highly contested process articulated by different middle-class consumption interests who use the local political system to protect their environmental positional goods (Hirsch 1976). These areas have also been subject to high levels of economic growth and development (see Murdoch and Marsden 1994) both in terms of inward investment and internal firm formation, through the development of new industrial units and the growth of a more diverse service sector as a consequence of the purchasing power of middle-class groups.

Second, the *contested countryside* is represented by those rural spaces which lie outside the core commuter catchments and, as yet, may be of no special environmental quality. Here farmers (as landowners) and development interests may still be politically dominant and are thus able to push through development proposals associated with agricultural diversification and small industrial schemes. These are increasingly opposed by the recent waves of incomers who adopt the positions that are effective in the *preserved countryside*. Thus, the development process is marked by increasing conflict between old and new groups (see Ward *et al.* 1995; Flynn and Lowe 1994).

Third, the *paternalistic countryside* refers to areas where large private estates and farms still dominate and the development process is decisively shaped by established landowners and farmers. Many of the large estate owners and farmers may be facing falling farm incomes and are thus searching for new sources of income. In this sense many are trying to convert agriculturally – productive capital into new forms of landed capital, by selling off redundant buildings or leasing land out to private farm management companies (Whatmore *et al.* 1990). Landowners and farmers are unlikely to be constrained in these strategies and they still take a long and custodial view of their land, property and village contexts, adopting a modified stewardship role in the rural community. These areas are less likely to be under great development pressure than either of the earlier two types.

Fourth, the *clientelist countryside* is likely to be found in the remote upland rural areas where agriculture and its associated political institutions still hold sway, but where farming can be sustained only by state subsidy, such as less favoured areas per capita payments and welfare transfers. Processes of rural development are dominated by farming, landowning, local capital and state agencies, usually working in close corporatist relationships. Farmers will be dependent on systems of direct agricultural and agri-environmental support (*e.g.* Environmental Sensitive Areas (ESA) payments), and any external investment is likely to be dominated by employment concerns and the welfare of the rural community. I detail the evolutionary nature of this type of countryside with reference to Wales in Chapter 6.

The typology is an attempt to characterise the processes of uneven rural change as driven by different sets of internal and external powerful interests. These in turn tend to 'create their own spaces', shaping rurality in different directions. In the *paternalistic countryside* one could expect landowners and farmers to still hold sway not only, as Newby *et al.* (1978) suggested, in demarcating and upholding cheap agricultural labour markets, but also in shaping the degree of agricultural diversification and the shift of assets between agricultural operations and other land development opportunities. Moreover, the arrival of the MacSharry reforms of 1992 in creating greater incentives for set-aside and environmental conservation, coincided with the devaluation of UK national currency (and the 'Green Pound') and the relative uplift in European regulated price support levels. This provided a fillip for the intensive producer, further creating the conditions for land concentration, intensive production and contract farming in many parts of arable Eastern England.

In addition, the patchy development of agri-environmental measures (such as farm stewardship and Environmentally Sensitive Area (ESA) schemes) in the intensively farmed areas has further recognised and reinforced the custodial property rights of farmers and landowners; with their traditional representative bodies (the National Farmers Union and the Country Landowners Association) now more eager to assimilate the post-productivist countryside as long as it is shaped around the exploitation of their members property assets (see below). In this sense then, the significance of the landed classes in the *paternalistic countryside*, has been far from a static phenomenon in the post-productivist era. Indeed, partly because of the incremental and partial nature of many of the policy reforms (*i.e.* failing to really address sustainability, and upholding agricultural freehold property rights), they have been able to maintain their local and regional grip in many of the most intensively farmed (and thus potentially unsustainable) bits of rural Britain. Agrarian paternalism has then adapted to both local and external demands by accommodating the broader environmental stewardship of agricultural land. It is now reproduced around both an agricultural and environmental base. And, with the current absence of a rump of preservationist middle-class, the degree to which they can truly deliver the alleged benefits of this broadened stewardship role remains largely unobserved at the local level.

In the *clientelistic* areas of upland marginal Britain, where rural economic development concerns sit side-by-side growing formal environmental protection and continued agricultural state support as a multi-faceted system of public welfare from Whitehall and Brussels, it is the growing dominance of the representatives of the 'provider' agencies which begin to shape developments. Also, these areas have traditionally been the sources of extractive industries (*e.g.* coal, water, energy), and with their physical remoteness and weak labour markets, they have been prone to the relocation of waste, and toxic and defence industries. The recent privatisation of most of these utilities in Britain has also tended to increase reliance upon the private providers of jobs in these areas. As a result, for instance in Wales and in parts of Cumbria, one now sees the emergence of some interesting juxtapositions of private and public *clientelism*, where National Park Authorities, mining firms and public and semi-privatised development agencies (such as English Partnerships), play influential roles in shaping developments. The relationships and dependencies between the provision of jobs and the provision of consumption spaces for amenity comes into stark relief (Munton 1995). And as a result of the weak labour markets and the dependencies these bring in terms of the powers of inward investors to bargain with public authorities, it means that the outcomes for environmental care and plan-making are of a different order and shape than in the other types of countryside developed here.

Such emerging ideal types begin to capture the processes of differentiation and dispersal occurring in the post-productivist countryside (see Table 5.1). They require further development and analysis. In considering such expressions of rural change it is also necessary to explore the unifying theoretical concepts which help to provide a more coherent picture of this rural differentiation. That is to begin to understand the causes of difference. In what follows, an outline of some of the key common dynamics needed in approaching an understanding of contemporary economic change in the rural sphere is developed. Collectively, these conceptual parameters begin the shape an approach to the study of the rural economy, after the rural and agricultural crisis of confidence of the past 15 years. The question which pervades this discussion is to what extent does this add up to a coherent picture for the contemporary rural economy? Does it suggest the demise of national rural structures and the differential emergence of regional ruralities? What sorts of concepts do we require to begin to answer such questions?

Conceptual dimensions of post-productionist nature

The new centrality of rural space in regional and national economic terms derives at a general level from the changing bases in which economic relationships are *formed*. They are regulated by actors and agencies through social and cultural interactions, using different combinations of knowledges, artefacts, attributes, technologies and natural phenomena (Law 1992). A host of regulationist and social constructivist literature in economics and economic geography has highlighted how market relations are socially and politically constructed and embedded in institutions, norms of behaviour and trust relationships (see Grabher 1993; Salais and Storper 1992; Amin and Thrift 1995). Economic restructuring or uneven development thus has a social and cultural basis both in general and particularly in the rural domain (see Marsden *et al.* 1996b). This needs to consider the nature and the quality of economic and power *relations* (see Figure 5.1).

Table 5.1 Dispersal and differentiation: the statistical construction of the post-productivist countryside

Demographic

- Up to a quarter of the English population lives in rural districts.
- The population in rural districts grew faster (24%) than in England as a whole (6%) over the last 25 years (1971–1996).
- People in rural areas tend to be older than in the rest of the country: 18% of the rural population is aged over 65 years compared with 16% nationally.

Economic

- 29% of the English labour force works in rural districts.
- Employment growth in rural areas has been faster than in urban areas.
- Unemployment in rural districts is generally lower than average (4.2% for rural districts and 6.1% across England), but there is significant variation between areas, with some areas well above the national average.
- The mix of industrial sectors in rural areas is similar to urban areas – the main difference is agriculture, which accounts for 4% of employment in rural areas and less than 1.5% in urban areas.
- There is variation within rural areas, however, and agricultural employment can be as high as 20% in some rural districts.
- 8 out of 10 counties with the lowest GDP per capita are rural.
- Agriculture's share of national income has fallen from 2.9% in 1970 to 1.0% in 1998.

Environment

- Rural districts account for 82% of the total land mass of England.
- Over half the national population would like to live in rural areas because of the perceived quality of life they offer.
- The rural environment is under threat from economic development and agricultural intensification:
    - in 1993, 53 km^2 of rural land – an area the size of Oxford – was developed (41% of all land developed in England;
    - one third of all hedgerows in England were lost between the 1984 and 1993 (and these losses follow dramatic losses over the 1960s and 1970s).

Social

- Despite low unemployment, poverty exists in rural areas; though levels are slightly lower in rural areas (30% of the population) than urban areas (40%).
- DETR studies show poverty in rural areas can be a persistent problem.
- More people in rural areas rely on private transport – usually cars.

Figure 5.1 Dimensions of the social and political economy of rural development

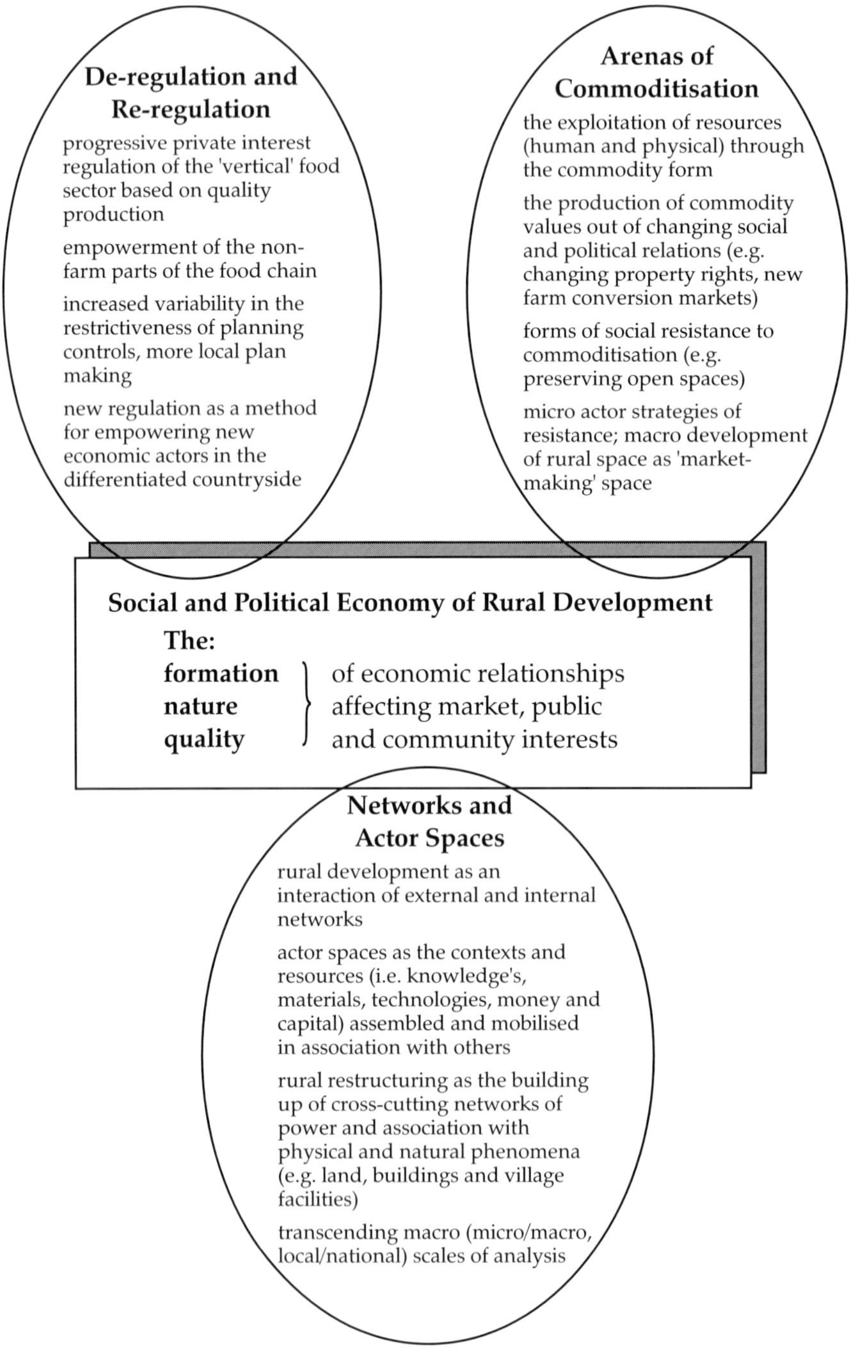

This social economy perspective holds at least three components. First, it focuses not only on the patterns of uneven rural development but also on the ways in which combinations of market, public and community interests and networks *carry* the process of rural development forward. This is the social dynamic of rural restructuring. Second, it is also necessary to examine the ways in which these differentiating rural trajectories (such as those in the typology above) actively redefine combinations of local rural resources in new ways. This has, as will be outlined below, been quite firmly associated with the *commoditisation* dynamic, whereby rural resources have been attributed a market value for exchange. For instance, former agricultural resources (barns, housing, land, woods) are being unevenly exploited by new groups of actors; in addition, new rural amenity and niche products are being created through the reconstitution of place and identity. Third, a focus needs to be placed on the ways in which traditionally perceived economic relations become embedded and carried through different sets of social, political and regulatory actors and agencies. For instance, ESA payments and structures funding are regulated through the Ministry of Agriculture, Fisheries and Food, (MAFF) down to farmers, and the Rural Development Commission in England allocates grant funding to particular areas over others. Increasingly, the institutional frameworks carrying economic benefits to rural areas are highly spatially variable.

The term 'social economy' connotes the integration of these factors of dynamic rural change. This grounds the multiplicity of economic relations in their social and geographical setting (i.e. methodological situationalism). It provides a theoretical basis for developing a systematic comparative approach to the new post-productivist rural diversity. This deals with rural space as a holistic entity. Such an approach is particularly apposite to apply to rural conditions, given the growing and diverse demands placed on rural resources and facilities and the degree to which these relate to different conceptions of quality of products, services and economic relationships. Providing quality (often with specific reference to place of origin as an authentic locale) becomes a way of market entry and maintenance for many rural economic sectors. This relative success then influences overall development. Of course, not all areas benefit from these engagements in new exchange relations. Indeed, almost by definition one can expect greater local and regional inequalities to be generated as less emphasis is placed upon a national system of social and economic provision. What are some of the key parameters of this social economy?

Deregulation and re-regulation

The political economy of the 1980s and 1990s was dominated with the attempts by government to deregulate market relations and to reduce state burdens by privatising former state assets. This process has not slackened in the 1990s, with government attempting to reduce interference in economic activity and to restructure the institutions, including government departments and rural-based agencies. These processes have served to provide different competitive advantages and disadvantages to rural spaces. On the one hand, the agricultural corporatism which had dominated thinking on rural development over the post-war period has been undermined by the necessity of reducing financial subsidy, eliminate marketing boards and wages boards and encourage 'free trade'. In terms of the vertical food sector this has acted to empower the non-farm parts of the food chain (particularly retailers and processors). Through the development of new food safety legislation from Whitehall (e.g. the Food Safety Act 1990) and Brussels (e.g. the Hygiene of Foodstuffs Directive 1995) retailers have been awarded new powers to police the food sector by essentially private means (see Marsden and Wrigley 1996). The changes have led not so much to de-regulation as to new forms of *private interest regulation*, whereby, private sector actors and agencies take responsibility for the successful implementation of policy ostensibly in the public as well as private interest (Marsden and Wrigley 1995).

On the other hand, in terms of the non-agricultural related restructuring, early attempts to deregulate the planning system were vouchsafed by the rural middle classes keen to protect their increasingly positional grip on large chunks of 'middle England'. This subsequently led in the 1990s to the uneven development of local and regional power coalitions of interest which reinforced local plan-making along conservationist lines. The Planning and Compensation Act (1991) determined that development control decisions should be made in accordance with the plan unless material conditions indicate otherwise, following the extension of local planning to all rural land under the 1990 Town and Country Planning Act. The net effect, despite over a decade of neo-conservative philosophy, has been for a more not less regulated countryside, and one which is subject to capture by particular local residential, producer and consumer interests within different spatial contexts.

More complex models of interest representation have developed over rural space which are sensitive to local social and economic conditions. For instance, agricultural changes in some regions have become more reliant upon the public-interest local planning system, such that farmers (in the buoyant growth regions like the south east) are obliged to negotiate changes through the development control system. Passage through these systems of mediation are not just associated with political gains and losses. They moderate and regulate the flows of mobile and fixed capital into and out of rural space. They are carried by networks of actors who are differentially empowered by these regulatory changes. In many parts of the *preserved* countryside in South East England, and increasingly in the *contested* spaces of the South West, farmers become the bystanders to the powerful coalitions of middle-class fractions and non-agricultural entrepreneurs who have gained more than simply a physical foothold in rural space. Their power and legitimacy have grown in the face of both agricultural retrenchment and the onset of local planning systems which embrace middle-class protectionism in the face of an ideological questioning of planning regulation within the free-market orthodoxy.

Theoretically then, as with subsequent parameters, recent tendencies have been far from unilinear or cast simply and strictly along economistic lines. The new post-productivist countryside, a terrain where consumption and exchange vie seriously with the more traditional productive activities, becomes a contested social milieu, through which the mobility and fixity of capital occurs. The regulatory diversity, and some would argue fragmentation, has most recently been recognised with the publication of White Papers for England, Wales and Scotland. These explicitly admit that government cannot, even if it now so desired, clearly and purposely regulate for the countryside in the traditional sense. Rather, as a result of the deregulation-re-regulation dialectic of the past 15 years, all it can do is to give a voice to the sanctity of rural life and to rely upon the rural economy to prosper from its own resources, both physical and human.

Hence, the regulation-deregulation dynamic provides a very uneven playing field for economic activity in rural space. It changes the sources of competitive advantage and disadvantage at a locale's disposal. The different countrysides which constitute modern rural space are having to develop their own regulatory and institutional frameworks. For instance, a growing distinctiveness can be observed between the institutional frameworks operating between the different countrysides outlined in the ideal types above. This suggests the need to develop theoretical and conceptual constructs concerning what Clark (1992) has called 'real regulation' (see also Munton 1995; Flynn and Marsden 1995). It is necessary to understand more fully how regulatory activity and authority becomes *spatialised*, and how these spatialisations begin to influence the flow, direction and density of economic relations.

Arenas of commoditisation and social resistance

Whilst by no means the most inviting of terms, commoditisation is a concept which is central to the understanding of contemporary rural restructuring. The complex process of socially exploiting resources (particularly labour, but also increasingly land, nature and the built form of the countryside) through the extension of the marketised commodity form has its more recent origins in the sociology of agriculture literature. Here, following a conventional Marxist interpretation of the reasons for the maintenance of family farming and the progressive expansion of capitalist agriculture and agribusiness, the concept can now be more imaginatively applied to an actor-oriented perspective for rural restructuring. Commoditisation represents a variety of social and political processes by which commodity values are constructed and attributed to – in this case – rural and agricultural objects, artefacts and people. It does not represent one all-encompassing process which, for instance, transforms agricultural labour processes or simply puts a price tag on a formerly publicly – owned piece of property. Rather, it is important to see it as a diversely constructed phenomenon around which rural development processes coalesce and then diffuse. The question it poses in the complexity and diversity of contemporary rural space is ' how do commodity and other social values shape social practices?' and 'how are commodity relations and values generated, and challenged, through the active strategising, network building and knowledge construction of particular producers, consumers and other relevant users of rural resources?' It involves exploring how people, objects, and combinations of natural and human artefacts are actively assembled in rural places, and then how these are used and valued in market exchange. Conversely, it is also important to understand how some actions remain outside the commodity form. Many local action groups in rural areas assemble resources in order to fight these 'commoditisation battles' with developers and planners. Murdoch and Marsden (1994) have documented at length how in the *preserved* countryside of the South East these middle-class groups have held a strong influence in shaping the social and physical spaces in which they reside.

Below this generalised level of discussion concerning commoditisation, however, it is necessary to examine in more specific detail how social actors assemble and construct 'value' in rural space, and then assess the degree to which this seems to take on a commodity form. Explicit struggles may occur over the attribution of social meanings and values, access to resources and in relation to issues and differences of social identity. Hence to explore commoditisation processes is to examine the means and strategies by which social resistance occurs around prevailing commodity definitions and practices. As with the discussion of regulation above it is by no means a predictable process. Rather, it highlights the socially active nature of what, at a distance, might be seen as economic relations. Hence, it is a central element in the changing social economy of rural space.

In some long-standing empirical work on rural areas in the UK (see Marsden *et al.* 1993; Murdoch and Marsden 1994; Marsden *et al.* 1996b) the multidimensionality of this process for different rural spaces has begun to be documented. This provides a useful comparative device in understanding rural differentiation. The commoditisation of redundant farm buildings, for example (see Kneale *et al.* 1991), was highly spatially variable in the late 1980s. Similarly, the development of recreational sites, such as golf courses, tended to cluster in the *preserved* countryside of south-east England. This re-commoditisation of former agricultural land is differentially regulated by local planning systems such that high levels of (market based) development pressure can be resisted in the preserved countryside and, through effective developmental planning in the poorer rural regions of the uplands and the north can, with considerable difficulty, be attracted. This demonstrates that developmental processes in the post-productivist countryside are far from simply reliant upon the uneven development of capital and economic activity. If this was the case very different rural and economic landscapes would be now witnessed. Rather, people and their institutions have to 'carry' and resist the commoditisation processes. And the balances between resistance and accommodation help to shape rural space both economically and physically.

More work needs to be conducted on the ways in which value is constructed in the countryside and the methods used by distant and local actors in generating and resisting change (see Harvey 1996). The past 15 years have been marked by the development of new 'market making' in rural areas, stimulated by deregulation and boom and bust economic cycles, not least in the housing sector. These processes have been a major agent of change for rural space, continually forcing productivist agriculture onto the back foot. The continuity and succession of family farming has come under particular threat, as farmers, and particularly their sons and daughters, found it more difficult to resist the attractiveness of seeing their land and buildings as multi-faceted capital assets rather than as productive assets. With the emphasis on maintaining the mobility of capital and reducing 'sunk costs' and the risks to investment, there is greater emphasis on flexibility of ownership and innovative forms of property relations (e.g. options, short-term leases and contracts, novel and divisible rights).

Many of the traditional rural interest groups have eventually conceded to these processes. For instance, the Country Landowners Association, eager to protect their members' freehold property rights in the face of the demise of the agricultural priority and the growth in demand for rural amenity, now actively promotes the concept of 'rural enterprise zones'; areas like their urban counterparts, which are free from planning control and provide entrepreneurs with developmental incentives. More generally, the recent batch of Rural White Papers (and particularly the English version) constantly reminds the reader of the necessity of equating good countryside with a working countryside. It is important to recognise, however, that the types and the general architecture of the market-making ruralities which have developed over the contemporary period have the common feature of using *particular* commodity-based methods to deliver rural development. In addition, and by no means unrelated, these developments have tended to be privatised. That is they

concentrating on the formation of new food 'circuits' are demonstrating the significance of embedded social networks as a forerunner for local rural economic development (Picchi 1994; Fanfani 1994; Iacoponi *et al.* 1995). Indeed, as the Common Agricultural Policy increasingly becomes cornered by trade liberalisation tendencies on the one hand and pressures to de-intensify agriculture on the other, more reliance will be placed upon assuming a positive relationship from the existence of dense social and economic networks and economic development gains in line with the somewhat mythologised 'Third Italy' model. This is already being promoted by EU Structures funding which expects 'partnerships' and locally developed networks to be in place as animators of local economic development.

The reality is obviously much more complex and variable and we should be cautious in assuming that the 'density' or richness of extant networks necessarily leads to positive economic growth trajectories. Nevertheless, a network and actor-space approach, if developed though intensive research, does have the capability of progressing an understanding of rural social and economic change, as recent commentators have concluded (see Murdoch and Morgan 1996; Van der Ploeg and Saccomandi 1995). It also, interestingly, begins to expose rural areas as potential sites for economic innovation and new forms of exchange and transaction, associated potentially with more sustainable food and environmental systems. This gives new forms of rural development an infusion of contingency and possibility which they have for so long seemed to lack. Moreover, from an analytical point of view, the focus on the content, power and contingency of network construction does not necessarily imply a populistic retreat to localism in the face of the ravages of harsh global forces. As many scholars, including social anthropologists are aware, it is through a more sophisticated analysis of the 'local' that a broader comparative analysis of capitalism and globalisation can be built. One does not exclude the other. For instance, in a telling passage Miller (1995) has recently argued in a discussion on globalisation:

> *'such ethnographic observations are vital if terms such as 'post -Fordism' are ever to be more than glib generalities. Our micro-studies of consumption are not a retreat from political economy because we are finding that the local has become the commanding heights of the political economy. It is in here that we can relate directly to questions of, for example the comparative experience of World Bank sponsored structural adjustment'* (p. 10).

Differentiation and coherence in rural change

Earlier approaches in the political economy tradition in the 1980s placed great emphasis upon understanding agriculture as a sub-set of the international food system, with rural development as an adjunct to these processes (see Goodman and Watts 1994; Marsden *et al.* 1996b). Since that period, scholars have begun to appreciate the growing centrality of rural space after the demise of the post-war compromise and to begin to tackle the diversity of rural space from a more nuanced social and political economy perspective. This chapter has examined some of the conceptual dimensions necessary to progress this development. They are dimensions which open the analysis of rural space to broader social science endeavours associated with the questions of the state, consumption and the possibilities for sustainable forms of production and consumption. Such progress needs to be founded on a process of carefully integrating aspects of nature and consumption into analyses. While this may be easy to suggest, it demands further conceptual refinement and specification.

Over the recent past many rural areas in Britain, as well as many parts of mainland Europe, have seen considerable growth in new employment and capital investment. Higher rates of new firm formation have occurred particularly in the more accessible rural areas, compared with urban areas. Such economic generalities serve, however, to mask the variable and socially-active forces which contribute to these aggregated changes. The three key conceptual dynamics outlined in the second half of the chapter (regulation, commoditisation, and actor spaces) give some theoretical pointers to the ways in which more effective comparative analyses of rural economic change can be progressed, both within nation states and internationally. Such an endeavour becomes all the more important given the differentiation currently occurring in the development trajectories of rural space. Under the 'post-productionist' conditions now prevailing, and, moreover, likely to continue, it is particularly relevant to consider how social, political and economic relations and outcomes *become spatialised.* Critical here is to appreciate how economic connections, distances and territorial arrangements are constructed; how they are made and become active ingredients in the dynamics which help to create the new rural differences. The next two chapters in this section explore what these post-productionist perspectives might mean for particular rural regions (Chapter 6) and for the redefinition of agriculture in the post-productionist rural space (Chapter 7).

6 Clientelism in the Welsh countryside

Introduction

Despite the continuance of both agricultural productivism and post-productivism in different parts of the British countryside, in some regions both of these dynamics have been less clearly expressed. Many of the 'remoter' rural areas in the upland agricultural regions have long suffered the (agriculturally-productivist) stigma of being defined as 'marginal – agricultural regions', in that they suffer from the post-war problem of 'agricultural adjustment' in realigning their farm structures to capture scale economies. Also, and for completely different reasons, these areas have not faced the powerful surges of counter-urbanisation so relevant in the 'preserved' and now the 'contested' countrysides. This 'vacuum' has been filled, however, partly as a consequence, by a stronger state welfarist intervention associated both with agricultural corporatism (based on farm family survivalism rather than productivism per se) and, as we shall see, more recently the emergence of a rural development dynamic largely associated with the non-agricultural rural economy and land base.

These socio-political conditions represent a distinct type of differentiating countryside: what we have termed; 'the Clientelistic Countryside'. This chapter will outline the development of this type of countryside with reference to rural Wales. First, I will outline the post-war socio-historical conditions which have established this particular rural development trajectory, before, second, seeing how, under the more recent conditions of agricultural crisis, clientelism struggles to take on a more rural development rather than strictly agricultural form. The final part of the chapter traces through what this transition in the Clientelistic countryside means for overall rural change in Wales.

In the previous chapter, I defined the clientelistic countryside as belonging, by and large, to the remoter rural areas where agricultural corporatism in its traditional productivist sense still retains a considerable hold. These areas have only marginally experienced the social transformation arising from the forces of counter-urbanisation, and the social welfarist model of agricultural support and rural development is still prevalent (Smith 1992; Murdoch 1992). Despite over two decades of Conservative government administration which attempted to dismantle corporatist relations and rural welfarist policies, these areas have remained dominated by clientelistic relations between the interests of farming, landowning and local capital on the one hand, and state agencies involved in promoting primary production or rural development on the other.

Systems of direct agricultural support, before the current piecemeal shifts in CAP (e.g. Less Favoured Areas policy, sheep meat subsidies and Hill Livestock Compensatory Allowances) have continued to encourage the restructuring and 'modernisation' of traditional farming, even though local sympathies towards income support rather than price support are frequently expressed. Although pluri-activity is a well established business strategy for farmers, the scope for additional, locally generated economic diversification is very limited, and what there is usually depends upon state support and/or external capital. Large-scale private and public capital may be attracted to such localities in order to exploit their natural resources and pockets of surplus labour, and the damage caused by such development projects, agricultural modernisation schemes or afforestation is often muted and voiced most strongly from outside the region. Local politics is often dominated by a concern for the welfare of the 'declining rural community' (e.g. job creation, social housing etc) and an ambivalence about external finance but nevertheless a strong dependence upon it.

While these represent some of the key general features of the clientelistic countryside, it is important to recognise that they are far from static and that they have evolved through a period when state policy generally, and agricultural and rural policy in particular, have significantly changed from their earlier post-war certainties. The following analysis will outline the development of the clientelistic countryside as it relates to one rural region-Wales. It will demonstrate how the rural land resource in this rural region is being re-defined around distinctive clientelistic lines, where the State, increasingly, at a regional level plays a dominant role; and where the basic tenets of agricultural productivism become significantly challenged if not completely overcome. Indeed, most recently what we see is the contested evolution of clientelism as the social-economic forces impinging on such rural regions begin to change

Coping with marginalisation: the evolution of clientelism and the quest for identity

Rural Wales has always been difficult to define in any coherent way. As Williams (1983) and Murdoch (1992) recognise, the definition of a coherent Welsh economy is significantly hindered by a range of East West links between England and Wales, with little connection operating between the North and the South. Similarly the pattern of farming has traditionally varied according to variable terrain and the different access to markets; with North–west Wales traditionally concentrating upon butter and cheese manufacture for the nearby urban markets (mainly in England); with the mountain areas of north and central Wales predominantly given over the beef and sheep production. In the more lowland southern and eastern areas there is an emphasis upon arable crops and livestock fattening together with the production of butter and cheese (Howell 1977).

In terms of population some areas (notably in the east and the south) have recently experienced population growth through in-migration. Whilst this has occurred in other areas to varying degrees, many of the more peripheral rural areas in mid and west Wales continue to experience out-migration, particularly of the younger and economically active age groups. Although rural Wales has actually experienced an overall increase in population since 1981, this is directly related to the in-migration of an older generation and not necessarily an increase in the birth rate within the region.

The region is. traditionally characterised by 'close-knit' communities many of which are under considerable threat from population decline, out-migration and a lack of economic diversity. The indigenous population have in most cases strong traditional family ties in rural Wales. This is reinforced by the concentrations of Welsh speakers. It is a population which, however, has to increasingly rely upon its own resources, with an over dependence upon declining primary industries and a lack of opportunity for small companies to develop and to be able to compete with larger firms. There is a significantly lower proportion of service sector industries setting up in the region compared to the lowland areas of England. And, regional GNP has been persistently below the national average. The main employers in Rural Wales are the public sector, manufacturing, distribution and hotels and restaurants.

Although agriculture only accounts for around 9 per cent of the working population of Wales, it remains dominant in the lives of many people living in rural Wales. As I shall discuss below, agriculture has experienced significant falls in incomes in the 1990s following an already large-scale fall out in the number of farm occupiers up until that decade.

While these problems have been exacerbated in the 1990s, (not least because of the decline and centralisation in public services and public sector employment), they are much more longstanding. They partly arise from and are reinforced by at least two phases of rural clientelism; one which stems from the agricultural productivist period (post war 1940s-1980's) and the more recent post-productivist period. I shall deal with these in turn, for they both express the central thrust of rural development which governs rural Wales.

Productivist clientelism: the definition of rural marginality

The gradual disappearance of a landowning class in Wales during the first three decades of the twentieth century presaged the growth of a traditional and resilient family farm structure in rural Wales. Between 1918 and 1922 it is alleged that every major Welsh landowner placed at least part of their estate on the market (see Davies 1974). The growth in owner occupation was complemented by the almost complete erasure of subsistence farming and, more importantly for the future trajectory of rural Wales, the incipient incorporation of the owner farmers into state-supported productivist agriculture. As Ashby and Evans (1944) began to point out, farming was becoming more specialised and mechanised, and the relatively small scale of the Welsh owner-farmers steadily began to be seen as a major restraint upon their success. They were increasingly playing an 'economies of scale' productivist game they couldn't win, but were, nevertheless, obliged to participate in as part of the national agricultural strategy. One consequence of this process was the virtual elimination of wage labour from the countryside whilst the number of independent farmers stayed constant: 'leading to the further simplification of the agricultural class structure to a mass of family farmers' (Day *et al.* 1989).

These conditions set in train not only an agricultural policy which was to progressively tie the Welsh farmer (and his land) to the UK wide national strategy of productivism and state-led industrialised production, but also, by dint of the its consequences of mechanisation, we see the social effects of this in terms of rural agricultural depopulation. Whereas in the 'preserved' and to a lesser extent 'contested' countrysides we already at least from the 1950s, have the steady ex-urban expansion of population of many rural villages, in Wales we see rural areas losing population disproportionately by dint of the national agricultural strategy and the absence of counter-urbanisation.

It is the very duality of these trends which begin to define rural Wales as a 'problem region', both in the agricultural sense as a region of small farms which are struggling to modernise under the economies of scale orthodoxy of the day; and second, through the growing rural development problems this creates in terms of rural depopulation in the absence of counter-urbanisation. These trends in rural Wales represented a silent rural revolution (see Murdoch 1988) but they also set the conditions for what we might term an *arrested marginalisation* of upland rural regions, whereby such areas were defined increasingly as marginal and state support was needed distinctively to address this marginalisation. This created the seeds for the clientelistic rural policy associated first with addressing the needs of agricultural modernisation and second those of rural deprivation and depopulation more generally.

Whilst by 1941 we see that the proportion of owner occupation had largely remained unchanged for at least a couple of decades (at 37 per cent, Williams 1985); between 1939 and 1972-the golden age of national agricultural productivism – we see a decline in the number of holdings by 44 per cent, with 61 per cent of farmers now being owner-occupiers. Also, between 1939 and 1972 the number of farm workers fell by 160 per cent, with the number of tractors increasing by 236 per cent. Despite the growing recognition of the wider rural development implications of such policies the precedence given to agricultural modernisation as the panacea for solving all rural problems developed a longstanding legitimacy in these emerging clientelistic socio-political conditions.
For instance, in 1956 the governments response to the Mid-Wales Investigation report (a report commissioned to examine the rural problems of rural mid Wales) stated that:

> '*The government believe that the creation of stable conditions in rural areas can be achieved only by encouraging the trend towards more stable and economic units and share the view . . . that financial assistance should forward this inevitable development*' (Arton Wilson Committee 1956; my emphasis).

This conflation between agricultural productivism and wider rural development policy continues into the 1960s and 70's with repeated calls for rural development assistance. However, by 1962 we still hear the celebratory ring of:

> '*more food with less labour. This is something to be proud of – increased efficiency leading to the production of more food with less labour. It is no good calling this the drift from the land. It is increased efficiency in agriculture*' (HoC Grand Comm. 1962).

and, further in justifying such an approach in terms of rural development problems, the Secretary of State, Cledwyn Hughes, could still argue in 1968 that:

> '*the decline in the number of farms and hence the number of farming families has been going on for many years, and we must recognise that before we can put a stop to the movement out of mid-Wales.... The economic base of the area must be sound. This is quite impossible with the present farm structure and if we are to raise the standard of living generally in mid-Wales we must have an increased proportion of economic farms*' (HoC Grand Comm. 1968).

The Beacham Committee in the early 1960s were to lay the foundations for a broader rural development approach associated with attacking the problem of depopulation and fragmented rural communities through advocating the restructuring of the settlement pattern (consolidation) and through the expansion of the economic base by encouraging inward investment (initially through large-scale construction projects). However, this occurred very much outside of the more powerful agricultural policy realm, by blaming the problem largely upon indigenous factors and conveniently ignoring the part played by agricultural modernisation. Help and assistance were seen to be increasingly needed in the 'problem region of rural Wales', but the clientelism which developed was not allowed to stifle the dominance of the agricultural modernisation paradigm of the time. The agricultural lobby could claim the sanctity of economic efficiency of agricultural units in contributing to overall rural economic development, while the rural development lobby has to be content with very much picking up the pieces left behind in its wake.

These originating factors of rural (productivist) clientelism lead gradually to the setting up of a range of key policy institutions which are quite unique. The Development Board for Rural Wales (DBRW) is probably the best example as a specific development agency which represented the need of the State to take concerted action in the areas of non-agricultural rural regeneration. The role of rural Wales as a contributor to the national productivist strategy (and its resulting marginalisation) was not only restricted to agriculture. The other major national contender for rural land in Wales was the Forestry Commission. Again, paralleling agriculture we see the growth of an expansionist strategy in Wales which gave priority to timber production over rural economic development. Higher levels of mechanisation continually reduce the local labour needs with, as Harding argues:

> '*the additional employment provided by forestry only managing to make up some of the jobs lost in agriculture*' (1985, p. 185).

Despite reservations about continued afforestation, the government concluded in the early 1970s that the Commission should still plant up to 55,000 acres a year. It continued to concentrate on acquiring land in the upland areas of Scotland and Wales on the grounds that in these areas:

> '*population is declining and the expansion of forestry can bring considerable social and employment benefits*' (FC 1974).

By 1979, the Commission had planted 192,000 hectares of conifers in Wales and dedicated 37,400 to private woodland schemes. As Cledwyn Hughes had cautioned in 1968:

> '*Forestry has made and will continue to make a valuable contribution to the rural economy, but we should be wise to avoid exaggerating the scale of the opportunities it can offer for rural employment*' (HoC Grand Comm. 1968; see Murdoch 1988).

Both of these major land using sectors thus dominate post –war rural Wales. And as the priority for rural Wales to play its part in contributing to the productivist national strategy becomes more embedded, so too does the process of marginalisation and peripherality of the rural economy. Land development is effectively captured by highly strategic national interests, and it is around these interests that other rural actors and agencies have to perform.

The flip-side of these productivist and nationalised land development conditions, however, is the evolution of a clientelistic state structure which attempts to 'arrest' the rural marginalisation tendencies inherent in them.

Clientelism in crisis: the deepening farm crisis and the weakening of traditional clientelism

This quest for state welfare support is given greater poignancy in Wales given its ambiguous, but nevertheless growing national status and identity. It is another irony of Welsh rurality in the post -war period that just at the height of its incorporation into the UK wide policy becomes the time to spawn a fledgling need to create a distinctive Welsh identity about rural policy. Murdoch (1992) documents this succinctly; demonstrating how, from the mid 1950s, and in parallel to the dominance of the productivist imperatives outlined above, the vocal need for a distinctive Welsh, more devolved rural politics begins to take hold.

These tensions were always there; but they rear their head, paradoxically, at the same time as MAFF in London reaches its pinnacle of legitimate authority over the Welsh and wider UK rural land resource.

Three policy developments in the 1960s and 1970s were fundamental to the development of this process of clientelistic devolution. First, was the dual development of the two farmer representative organisations in Wales (the NFU and the FUW); the latter which, in the context of the growing need to represent the case of the 'Small farmer', and to gain as much as possible from MAFF in terms of 'agricultural adjustment' moneys (e.g. Hill Farming Act, FHDS, LFA/HCLA payments) becomes a more radical and Wales-based representative in the annual price review process. This split in the representation of farmers only added in the long term to the construction of marginality of the Welsh agricultural case, and the 'begging bowl' climate increasingly associated with MAFF and CAP support for the 'upland case'. The differences inherent between the two main unions (as we shall see below) tends to restrict a more distinctive rural policy for Wales; giving more credence to the need for 'compensation' for the beleaguered Welsh farmer.

Second, there is the more general process of constructing Welsh governance, associated with the setting up of the Secretary of State for Wales, the Welsh Office and WOAD. Here we see the first signs of the growing tensions associated with devolving generalised powers, on the one hand, but attempting to protect UK wide conformity in the sanctity of UK agricultural policy on the other. For the UK government with regard to agricultural policy this was particularly pertinent. What we see here are the beginnings of a long-term set of tensions – those associated with the national need to maintain conformity of purpose in productivist policy at the same time as dealing with the need to steadily devolve responsibilities for decision making to the Principality.

As Randall argues:

> '*...in those spheres of government activity, such as agriculture ... where there was no established tradition of jointly administering English and Welsh affairs, it was much more difficult – not least because of the attitude of civil servants themselves who objected to the concept of 'area' administration because it conflicted with their functional view of things*' (Randall 1972, p. 360).

Third, was the growth of a real, if fragmented rural development lobby which increasingly became anti-agricultural and positive towards rural community development. The setting up of the Welsh Development Agency (WDA), the DBRW and the activities of a growing myriad of other organisations, all with different relationships to the State, begin to represent and to articulate the operation of rural welfare need, largely despite rather than because of the dominant agricultural productionism. The operation of these rural development agencies, however, tend to reinforce, in many ways rural Wales as a low income and marginalised rural region.

In these ways, and partly as the unrecognised result of the very failures of agricultural (and to a lesser extent Forestry) policy we see the growth of a clientelistic countryside. It is one which, more so than in the other types of countryside outlined in Chapter 5, is beholden to the State for its definition, its dynamic and its distributive powers. By the end of the 1970s, in the dawn of the new crisis facing productivism and national agricultural pre-eminence, it is the socio-state relations which call the tune around which the rural land and people try to dance. To get change, to solve a problem, or to achieve anything, by the end of the 1970s, it is to the monolithic state (whether in Newtown, Cardiff or London) one needed to articulate with. Rural Wales, through its steady incorporation with productivist agricultural policy (and all its consequences) could only find ways out of its growing marginalisation by negotiating and accommodating with its very increasingly labyrinthine institutional structures. This represented the almost complete power of state based rural marginalisation.

It follows from this, therefore, that in the case of rural Wales, much more so than in our other types of rural area, we see the development of a distinctive socio-political formation based upon a particular definition of a *marginal regional rurality*; with, an increasingly devolved state apparatus as its main guardian. It is true that as in the other areas, that agricultural productivism was indeed just as strong an impulse in shaping the foregoing nature of rural land development, but in rural Wales, largely because of its lack of clear success, and the very persistence of rural deprivation as a 'rural-problem,' and corresponding rural regional identity, we see the public sector play a much more significant role in shaping the trajectory of rural space. Through the cumulative process of arrested, agricultural marginalisation and an increasingly devolved state given the responsibility to sweep up the pieces, what emerges is a highly complex form of rural clientelism. One which begins to falter as the very basis of agricultural productivism itself is increasingly challenged in the 1990s.

New roles for the state in a period of post-productivism. Clientelism in crisis

Throughout much the productivist period the agricultural interests in Wales represented something of a stable policy community. This involved WOAD, the NFU and the FUW. As Jordan and Richardson (1987) suggest:

> '*There is a natural tendency for the political system in Britain to encourage the formation of stable policy communities, one of the primary purposes of which is to achieve a negotiated and stable policy environment. The underlying value is that, whenever possible, outcomes should be negotiated between policy professionals with each participant fully aware of the needs and desires of fellow members of the policy community. This is not to suggest that there is an absence of conflict, but that conflict is, by agreement, kept within manageable bounds, if only by concentrating on issues which are susceptible to bargaining*' (p. 181).

The areas of susceptible bargaining were associated continuously with the annual round of price reviews, and in later years, as the farm crisis grew, the degree of state compensation needed to shore-up the farming population. The WOAD, the NFU and the FUW (see Murdoch 1992) formed such a stable, if conflictual community. As Murdoch foresees in 1992:

> '*However the indications are that this situation will no longer be so cosy. EC milk quotas marked a turning point in this respect. The crisis of agricultural policy (persistent surpluses and an ever expanding agricultural budget) forced radical action to curb expenditure.... Milk quotas would seem to be just the beginning. The whole approach to EC agricultural policy from the UK government's position is characterised by a desire to 'get back to the market', in effect to rein in state expenditure*' (p. 176).

These clientelistic set of relationships in a rather paradoxical sort of way become much more acute and embedded as the farm crisis continues. As the foregoing analysis demonstrates, they tend for a long time to shape the crisis itself. And, as this crisis – essentially a crisis of small upland farming regions – deepens the main actors, the Unions, government officials and the elected representatives, become more significant in the shaping and articulating of this particular type of differentiated rural region. In short, growing gaps open up between this region and others discussed.

> '*These representatives, therefore, do not simply represent the region: they are forced to calculate, negotiate and compromise on what the region might be and what the region might need. These calculations, negotiations and compromises take place within institutional structures which reach into the region from the centre. These structures are designed to transplant regional representations back to the centre where they can be translated into 'manageable' demands to be accommodated within the existing procedures of governance*' (Murdoch 1992).

The 1990s represent a period then of deepening crisis in the agricultural corporatist relations and in the plight of the farming economy. Earlier methods of constraining production and establishing a more 'post-productivist' stance in terms of government policy, become increasingly derailed in the 1990s, particularly in the late 1990s. Farmers in Wales become more rather than less reliant upon income support (representing more than actual derived incomes by the end of the decade). In this sense public support, itself in crisis – both in direction and in volume terms – becomes ironically more significant to the smaller, upland farmers in the UK ; but it does so when the traditional livestock markets also suffer a severe blow through the effects of the BSE and related food quality crises. In addition, unlike many of their larger, lowland counterparts in other parts of the country, the smaller livestock producers have difficulty in reaching for the increasingly extended economies of scale that the still productionist elements of the CAP encourage. In short, by the mid 1990s the true 'crisis of the middle' affects agricultural areas like Wales. They are still locked into a productionist game which they increasingly cannot win; and with more and more pronouncements by the UK government of the need to reduce CAP support, they face a growing need under these arrangements for more rather than less income support. This is particularly associated with the growing annual conflicts between the farmers unions and WOAD relating to the scale and direction of Hill and Livestock Compensatory Allowances (HCLA's) (see Midmore *et al.* 2001). More public support is pressed for only to increase stocking densities for sheep that are continuing to decrease in value. Even more extraordinary paradoxes begin to surround the beef trade with the near total collapse of the red meat markets following the 1996 recognition that human transference of CJD from BSE (see Marsden *et al.* 2000a). The historically stronger and specialised dairy sector also starts to fall victim both to the

for shifting these socio-political conditions to a new plateau. One which puts strictly farmer representation on the back foot; and continually eats away at the very basis of the earlier, stable productivist compromise.

However, what we witness here is something far different to the new post-productivist compromises established in the preserved countrysides of south-east England, or the more contested countrysides of rural Devon. It is not rural preservationism which is fuelling the new,or revised clientelistic countryside. Rather it is a reformulated type of *rural developmentalism*. One which had its origins in the 1950 and 60s in rural Wales, as we discuss above in relation to the early establishment of the Rural Development Board for Rural Wales. Now, however, it has a broader agenda and a stronger experience and policy framework (e.g. Cork and Agenda 2000) from which to draw its sustenance. Moreover, it does not deny the need for a supportive and clientelistic state. Rather, it expects a broader base of support for: (i) strengthening rural communities; (ii) improving access to rural services (iii) spreading prosperity and evening out the problems of uneven development; (iv) developing rural skills and learning(v) investing in rural infrastructure and (enhancing the rural environment.

As in the case of Scotland and some areas of upland rural England these new tenets of rural clientelism have been built upon the experiences and networks associated with the LEADER and associated community development experiments and the realisation that the state already pays for more of the jobs in rural areas than any other type of employer in these types of rural areas. The logic is therefore still one of state dependency, but now it is not just primarily associated with the support for agricultural production, or support generated from the smoke filled bargaining rooms of Cardiff and Whitehall government offices. In part, it is about using these government agencies to unlock more of the funding for rural development from Brussels (whether it is associated with the structures or CAP budgets).

Hence unlike the UK and Wales orientation of the previous period of Productivist clientelism, what we see here is one built more upon (as I argue in Chapter 4) a system of multi-level level governance, whereby the *articulation of networks* between the local regional,national and European policy arenas is the norm. This does not come without its problems and discontinuities,however. But at its heart is the articulation of specific rural development problems that rural Wales faces at the various levels, and the central assumption that state support will still remain central to the protection and enhancement of rural Wales in the more globalised and post-productivist world.

Differentiating ruralities in the clientelistic countryside

In contrast to the other differentiating countrysides discussed in the last chapter, we can see that the reconstituted state continues to play a central role at regional and national level, both in articulating the rural development problems and in allocating support (not least funding) to the most rurally deprived areas. It is particularly at the regional and national level (within Wales) where these battles and contestations are fought; and where different contests in the clientelistic state are played out between different spatial realities (e.g. between the Objective 2 area of Mid and east Wales with the Objective One areas of West and North Wales). These clientelistic relations create their own spatial rural politics which are built upon the spatial designations of that state in the first place. Indeed, spatial politics and the questions of territorial justice and injustice become a key characteristic of the new clientelistic rural state *because* of the crisis of resources generated, at least in part, by that state in delivering its allocative rights and powers. The state is seen as protecting and promoting some areas and sub-regions over others, and historically some sectors (particularly agriculture) over others. The state in turn has to continually respond to these pressures by allaying fears of asymmetrical spatial fixes. The dawn of the National Assembly, with its sixty elected members, has tended to heighten the tensions of spatial justice and territorial uneven development. Rurality becomes politicised in new ways. The former sectoral dominance of agricultural productionism begins to give way to a more complex territorial governance.

In the run up to the establishment for the National Assembly 1999 the Development Board for Rural Wales, which was responsible for a large chunk of Mid and West Wales, was finally wound up in favour of an 'all Wales' approach to rural development being consolidated into the WDA. From thenceforth the regional offices of the WDA have taken on some responsibility for rural development. Such an 'all Wales' approach has, however, been so far hard to achieve, not least because of the new boundaries associated with Objective One and Two structural funding, and the newly emerging 'Leader Plus' zones. These new clientelistic spatial fixes are still taking shape; and they are being built upon different sub-regional traditions of partnership and concerted actions for rural development. For instance, the 'Mid Wales Partnership', based spatially in the old DBRW zone, exhibits an established set of partnerships between the Training and Education Council (now ELWa), the local authorities, the agro-food partnership and the WDA regional office. These triumphs over integrative complexity are not always easy to achieve in other rural regions, however they do witness the dawn of a new spatial approach to rural development based upon partnership working of the main agencies. Below these key players are numerous (and often locally very effective) enterprise agencies and community groups. In some cases competition between potential partners can drive out collaboration.

In the new clientelistic countryside that now typifies rural Wales we see new spatial compromises being established between the competitive pressure to obtain blocks of funding (either from Cardiff or Brussels or both), and the need to build effective partnerships and cooperative arrangements in order to meet the conditions upon which such competitions can be entered. This is a far cry from the still extant but dwindling 'pipeline' funding for agricultural income support, which is still geared to individual farms rather than the regional areas they happen to belong to. Here we have then an interesting paradox which plays itself out spatially in the clientelistic countryside: that is the contradictory ways in which at least three different sub-regional income streams find there way into areas like mid-Wales. These include: (i) the allocation of large proportions of income support to farmers from the National Assembly regional offices – this takes place on an annual cycle and absorbs most of the time of the large staff of the Agriculture Department of the National Assembly; (ii) the allocation of economic development and agro-food finances to the WDA regional office; (iii) the highly significant public funding associated with the main public services (health, education, transport) which tend to dominate employment structures in such places.

These types of clientelistic countryside are now struggling to regain a new coherence based upon these new spatial fixes, after the post-war period of highly regulated sectoral allocation and policy. However, running through this shift, from a sectoral to a spatial politics, is, as we saw in Chapter 4, a conventional agricultural sector in deepening crisis despite its history of exceptional status and high levels of public financial support.

These spatial fixes also effect the growing proportion of agricultural land now under some form of agri-environmental designation (Ty Cymen and the new Tir Gorfal schemes as well as the ESAs). Even these, however, in the Welsh case, begin to take on a more rural development focus, even though their origins are associated with agricultural de-intensification (see Banks and Marsden, 2000).

The old rifts and cleavages between agricultural politics and a rural development politics are still there, but they get increasingly blurred as a more spatial state-based approach (including agri-environmental measures, and the agro-food strategy) begins to take hold and the regional state is increasingly called to account for the growing territorial disadvantages.

Currently, we see much of Welsh rural land criss-crossed with often multiple state-based, area designations associated with the new compromises and tensions between rural development and agricultural decline. The rural geography of Wales is outstandingly a regulatory geography. It is increasingly three dimensional rather than two-dimensional. That is, rural land is regulated not only by a sectoral agricultural and forestry focus, and a restrictive land-use planning/agri-environmental system which protects and contains development. Both of these dimensions have to now compromise with the third emergent dimension of spatial rural development. This means the socio-political matrix of clientelism is even more densely organised around different spatial scales and hierarchically organised actors.

In one sense this dense web of clientelism represents a type of rural space which is increasingly differentiating from other types of rural area (e.g the preserved or contested countrysides of south-east and south-west England). The socio-political dynamics are not static and there is a constant *quest for coherence* in the competing demands placed upon the clientelistic state. As in the *preserved* and the *contested* countrysides these tensions and conflicts are being played out on the back of the general decline of the politics of agricultural productivism. But in the case of Wales this post-productivism takes a completely different and contingent shape. It is one which displays and indeed encourages a strong role for the regional state in the context of multi-level rural governance, extending as it does from local civil society up to the operation of a newly created National Assembly.

It is a rural world which still sits between and 'old style' and a 'new style' of clientelistic politics, one where government officials, quangos, elected representatives guide and construct rural space through a welfarist concern for the allocation of highly competitive and scarce state resources. It is post-productionist but also pro-developmentalist in character. This leads, it would seem, to widening rather than narrowing forms of uneven development *within* this regional rural space, however much the state authorities project the need to reverse these trends. This is a central paradox in the socio-political dynamic of the current clientelistic countryside.

From sectoral agricultural corporatism to agro-food networks: the reconstitution of the sectoral and regional state

In this final section of this chapter we will examine how these processes have been exemplified in the agricultural sector. As we argued in Chapter 5, one of the key processes of governance operating over the past decade, and one which has certainly influenced, at least the recent tempo of rural and agricultural governance, has been the processes of social, economic and political regionalisation. Overall, and this is clear in the rural sphere as elsewhere, the old sectoral certainties of state intervention, based upon cascading hierarchies of officials and highly structured forms of political inter-mediation have begun to decline as an effective means of governance. The decline of agricultural corporatism – as described historically above – for the case of Wales is an interesting case in point. Because of the forces of globalisation on the one hand, and the 'hollowing out' of conventional hierarchical state practices on the other, there has been a tendency for the emergence of more 'networked' and coalitional frameworks to develop which represent and construct new 'environments for action' (not least the national Assembly itself). Central to this has been the recasting of the role of the State from being a formulator and deliverer of policy, to now more of an orchestrator or conductor of networks. Increasingly, while state agencies may arbitrate over policy, they can now only act, *in relation to others*. As both a cause and consequence, more 'multi-level' governance structures come into being. These are vertically interlinked through operating at different spatial scales (European, UK, Wales, sub-regional, local authority), but they also have to engage with a wide variety of external actors and networks (both state and non-state) at each of these levels.

Currently, then, while it is important to recognise that certain vestiges of the 'old' sectoral corporatist regime are still operating in the rural governance of Wales, so too, we can see the emergence of this more flexible and networked, mutli-level governance structure. Indeed, as we shall demonstrate below here with reference to agro-food policies, how the conflicts between these models of rural governance, in a clientelistic region like Wales, tend to play themselves out in the very conflicts over the direction of rural and agro-food policy.

While the BSE scares and problems were started a decade earlier, it wasn't until March 1996, when the admission came concerning the possibilities for transference to humans, that the Welsh livestock sectors markets were seriously affected. This re-ignited a longer running crisis of markets in the livestock sector and led to a quite dramatic fall in farm gate prices. The crisis, however, was also a turning point in two other significant ways. First, rank and file farmers, for the first significant occasion, and increasingly frustrated by the lack of progress from their farmers union bosses, began to take action into their own hands by blockading imports of Tesco hamburgers from Ireland. They began to adapt a more effective 'supply-chain approach' to the problems they faced as opposed to a more delegated and arms-length and corporatist 'smoke-filled room' approach. Second, and increasingly once the 'New-labour' government came to power in May 1997, the established governance structures began to place less emphasis upon upholding, renewing and relying upon the old productivist subsidy and compensatory system for alleviating these more deep-seated problems now facing the sector.

Rather than really considering a radical shake up of demand regulation in the rural and agricultural sectors, and coming at a time when the transition for the Welsh Office to the National Assembly was taking real effect (1999-2000), much more emphasis was placed upon developing, from a low base, a supply-chain remedy to the agricultural crisis. This was activated through the establishment of the agro-food strategy and agro-food partnership in 1999; and for the first time brought together senior officials from the Welsh office and those in the Welsh Development Agency. This was also supported by the relevant ministers of the day (e.g. Ron Davies). These developments ushered in a post-corporatist world of more flexible and networked articulation of agricultural interests, whereby, for instance, active 'working groups ' were established to set targets for the different sectors of red meat, dairy and organic sectors. These were constituted by a whole variety of private and public sector actors who operated across the whole of the supply chains, rather than with the agricultural base itself.

The WDA (a spatially-based development agency rather than a sectoral one) was given the function of implementing the strategy and the partnership, which together with taking on the whole of the (non-agricultural) rural development portfolio, following the abolition of the Development Board for Rural Wales (DBRW), gave them a new major stake in the regional economic governance of the Welsh countryside. By 2001, 9million pounds had been directed to the initiative, with support ranging from funding for local food initiatives and farmers markets, through to large-scale investment in the dairy processing and equipment of big companies. In addition, this development was able to capture newly created resources coming in through the EU structures funding streams (Objective 1, from 2000 onwards); and is beginning to capture modulated monies coming through the 'second pillar' of CAP (through the creation of the rural development regulation plan).

Several points need to be made about the shift in broader rural governance with respect to these trends. First, the growing realisation of the need for a more regionalist and networked approach to agricultural and rural governance comes at a time in Wales when the overall philosophy of devolutionism, in the UK, is being significantly progressed. In addition, in macro-economic terms this is tending to adopt more of an overall governance, rather than government stance; and one which is built upon aspects of sub-regional inclusiveness and partnership. Also, with the advent of new rounds of Structures funding (Objective 1), the advent of the 'second pillar'; and the moves to engage, hitherto, more industrially-based economic development agencies into the ambit of rural and agro-food developments, funding mechanisms themselves again (as with the productivist/ corporatist regime) become a major spur for the creation of more effective partnership working and rural and spatially-based project management (see Franklin, 2000; Morgan and Mungham, 2000; Marsden *et al.* 1999).

This represents a new regional-based and diverse coalition of interests in the rural and agro-food sphere, whose consensus in part stems from the realisation that there are severe limits to what the State can achieve beyond a facilitative and coordinative role. This process spreads out to the non-agricultural rural development sector as well, with, for instance, the fledgling creation of a 'National Rural Partnership' (NAW 1999a) and the coordination of new LEADER Plus funding through the WDA. In addition, more emphasis is placed upon the coordination and bottom -up activities of local enterprise agencies and local authorities (see NAW, 2001a), and the role of the WDA in acting as a major coordinator of these plethora of activities. This becomes somewhat uneasy territory both for the farmers unions – who regard these moves as something of a weakening of their monopoly position – and the officials of the WDA – who are somewhat reluctant to develop a distinctive approach to rural economic development, for instance, arguing that they treat all territorial features in a similar fashion.

We see however, partly spurred on by the significantly deepening crisis in upland farming, post BSE and FMD, the development of facilitative clientelistic rural governance, albeit from a low base. Devolution allows this new form of rural governance to be also justified by the opportunities of regional distinctiveness which exist increasingly across rural regions of the UK. The more the crisis in agriculture continues, with for instance, the closure of export markets and local abattoirs, import penetration, and the 'over-subsidisation' of lowland farmers, the more it fuels the political need to develop separate strategies for Wales, Scotland and the English regions. This is seen in the development of separate rural development plans, structure funding objectives and 'farming strategy' documents (see NAW, 2001b). These are now promoting the distinctiveness of Welsh rural life, and small-scale family farming through the need to compete on the basis of social, economic and environmental quality food production, rather than just on price. The seemingly never-ending crisis in the livestock and dairy sectors has even persuaded the reluctant farmers union leaders that the 'status quo is not an option'. The new vision statement recently launched by Carwyn Jones (the current NAW minister for Rural Affairs) this year represented, despite a lot of behind the scenes arguments, a growing consensus between the unions and the rest of the agro-food network, in that it embraced sustainable development principles and recognised that the days of pipeline funding for the agricultural sector (particularly on the basis of quantity of production) are numbered. As the former President of the NFU Cymru responded:

> *'I welcome the launch of the Farming Futures document not only for the positive suggestions it makes towards improving the farm family's technical and managerial skills, marketing initiatives, environmental opportunities and making the most of their assets as farms and individuals, but also the clear realisations that these benefits would be considerably greater if we were*

integrated with one another' (Quote published at launch of Farming futures 2001).

The realisation that the old corporatist regulatory system tended to erode such integration, either between farmers themselves, or between different actors in the various supply chains has now hit home. As a result more of the significant actors are realising the benefits of developing associational capacity and cooperative ventures. These may not have worked all that well in the past, or indeed, create new difficulties in the present, but they are increasingly seen as a way forward. However, we still have to recognise that the corporatist regime casts a very long shadow. The Farmers unions are still concerned about drawing ' a line in the sand' with regard to any future potential shift from 'Pillar one' to 'pillar two' (i.e. modulation), and set about lobbying government officials with regard to agri-environmental payments (under the former very successful, Tir Cymen, and the more recent Tir Gorfal schemes) in ways which are reminiscent of the former glory days of winning new battles with Cardiff and Whitehall in the annual price reviews. Now, however, they have been forced to recognise, particularly in the Welsh case, the growing political power of new rural development interests, the growing, and more respectable organic movement, as well as their own weakening electoral position. Farmers, themselves are now a much smaller constituency. At the same time, many more farmers and their families are integrated into the wider economy through part-time and pluriactivities (40 per cent). In many ways the aftermath of the farming crisis has left the farming unions very much the victims of a re-invigorated 'pincer movement'; with the down-stream food chain and networked interests on the one hand, and the broader non-agricultural rural development interests on the other surrounding their particular political concerns and relationships with government. This was very much in evidence in the constitution of the 'Farming Futures' expert group, and in the now regular deliberations of the National Assembly's Agriculture and Rural Development Committee, to which the Minister has to regularly report. The latter adopts both a scrutinising role of government policy and, more embryonically, a policy development role. Hence we can see that in the Welsh case, by dint of devolution, the changing institutional apparatus has tended to encourage a more networked governance framework.

Such trends, however, now also tend to expose the growing incongruities and contradictions implicit in 'Pillar 1'. The Welsh Assembly agriculture department still employ the bulk of their staff in doling out welfare payments (e.g. HCLA's) to farmers in different regional offices. Yet, in income terms, prior to foot and mouth outbreak, the average net farm income in Wales (2000/2001) was £4100. Farmers are still overwhelmingly reliant upon subsidy for this income, with direct Common Agricultural

Policy subsidies accounting for 420 per cent of the net farm income of the average farmer in Wales. Subsidised farming systems are no longer working even in the interests of their main recipients, never mind other rural and urban residents.

It would be wrong to deduce from this account of the transition, however, that it will inevitably lead to the further marginalisation of farming issues in the policy developments concerning rural Wales. Indeed, as I shall explore more fully in Section 3 of this book, unlike the post-productivist socio-political landscapes associated with the 'preserved' and the 'contested' countrysides of England, farming in Wales is now creating a new raison d'etre, now based upon its wider contribution to environmental, social and economic sustainability principles. Rural environmental agencies, like the Countryside Council for Wales are increasingly seen as bodies to manage this process in their administration of schemes (like Tir Gorfal). Hence, what we see here is the emergence of a new rural governance system which is still based upon a strongly clientelistic framework, but one now which is attempting to re-integrate agriculture into policies using different sets of objectives. The relevance of 'small scale family farming' as an icon for the maintenance of the social and linguistic fabric of isolated rural communities is now stronger than ever; even though the reality is that further reductions in the numbers of farms seem inevitable. Rural development policy – now steadily encapsulating agricultural policy frameworks – thus tends to re-position farming; as a part contribution to achieving more effective rural sustainability. If rural sustainability discourses are partly a metaphor in South-east England for further and intense forms of middle-class protectionism, then in the clientelistic countryside of Wales they are increasingly used for re-integrating small family farms into the social and economic fabric of rural community life; and indeed, by default, as almost an antidote to the growing counter-urbanisation processes also affecting many Welsh rural villages.

The socio-political transition, within the rural clientelistic state, has therefore started to reconstitute agricultural governance rather than completely marginalise it. It has also opened the door for new forms of innovation in the rural and agricultural sphere, which demonstrate that a larger number of farms are engaging with the rural development paradigm, rather than with the older productivist one. This process has undoubtedly been accelerated by the devolution process, coming as it does, at a time of the growing realisation of the regionalisation of rural restructuring processes.

7 Agriculture in post-productivist rural space

Introduction: reintegrating agriculture into the rural economy

This chapter considers the ways in which agriculture, as the main land-based rural activity in Europe, is finding itself subject to more complex and differentiated demands, and to more regionally – and locally-based regulatory frameworks. It addresses the following question: How is agriculture coping with the post-productivist consumption countryside? These pressures are creating a bifurcation between the role of the farmer as a specialised producer, locked into distant markets and the role of the farmer as a small rural businessman or local environmental manager. It is likely that such a bifurcation will be enhanced by the pending changes to agricultural and structures policy, and it provides, as I show, a major dynamic in the differentiation of rural space more generally.

From a sustainable development perspective (or what I prefer, in conclusion to term ecological modernisation) this raises some important questions about the regulation of rural space and the integrative role that agriculture could play in this. Questions of diversity and integration are central to the development of new, more sustainable policy. In addition, as we have seen in the previous chapter, agriculture can no longer be viewed as a discrete sectoral activity or policy following its own productivist aims. Increasingly, as the main parts of this chapter point out, *it is how agricultural production interrelates and integrates with other types of development process which becomes important in conditioning rural development more generally*. Policies for rural sustainability need to incorporate this integrative potential of agriculture.

So far however, as I show through an exploration of some of the key parameters involved in the process of rural differentiation, academic and policy debates have failed to appreciate the degree of contradiction and lack of integration apparent in many of the development processes affecting rural areas in Europe. While there is much talk of viewing rural areas as 'holistic entities', which assumes some sort of natural condition of integration, again as we have seen in the previous chapter, we have to recognise that this is by no means inevitable. Indeed, as I argue in the conclusion, it will be a major challenge for rural sustainability policy to create such integration out of the fragmented, differentiated and contradictory ways in which combinations of state and market induced development has shaped contemporary rural space.

After outlining a perspective on rural differentiation in rural Europe, the chapter sets out some of the broad, and as yet rather under-researched, dynamics involved in shaping rural differentiation. These are divided into four land development spheres: *mass food markets, quality food markets, agricultural related changes and non-agricultural types of rural restructuring.* These attempt to focus not just upon the local rural effects but also on the regulation of markets and supply chains which can lie at various distances from local rural space. They are also subject to different types of regulation. The chapter then considers the implications of these uneven patterns of development for rural governance and regulation. In conclusion, I attempt to draw out the implications of these development trajectories and regulatory complexes in terms of rural sustainability policy.

Understanding why rural spaces are different

Globalisation, new technologies (not least information and communications technologies (ICT and the rise of GMOs) and the varied pressure for new forms of governance (such as regional devolution in the UK) have now more firmly denuded the role of strong nationally oriented and sectorally based rural and agricultural policy; and rural spatial differences rather than similarities become increasingly expressed. The question is how do researchers and policy-makers make sense of this turn to spatial difference? How can policies and governance structures be built which translate international processes in ways which protect the human, natural and physical asset structures of rural areas? This is made all the more difficult given the recent trends for rural areas to become progressively less self-sufficient and self-contained and ever more open to the wider forces (economic, social and political) shaping European rural development. This growing lack of self-reliance has not resulted in more uniformity across rural space; rather it is fuelling greater differences in structure, performance and social formation just at a time when many national governments, not least because of fiscal reasons, are looking to divest some of their control and responsibility to their rural heartlands. In this sense it is the regions within which rural areas are located that are the new regulatory focus for managing rural change.

Our analyses need therefore to explore the degree to which *different* rural areas are developing contrasting strategies of adjustment and compromise with both the state and the wider economy.

Of central importance here is the nature of the varying levels of regulation and their interactions with different social formations in regional rural spaces. It is this interaction which is central to exploring rural change. The interaction between types and levels of regulation and the degree to which social formations react to as well as influence these, is most clearly expressed in land development issues and processes. This provides an important analytical vehicle, therefore, in assessing rural differentiation and the influence and social shape of regulation and social formation.

Despite the decline in agricultural hegemony in many rural areas and the succeeding array of disparate political and institutional influences affecting the regulation and governance of rural space, we have to recognise that in terms of broader processes of restructuring in rural areas, and in terms of the consumption of rural resources, agricultural and broader land-based social and economic relations still have a significant hold on the shaping of regulation, and on *the processes by which rural areas are differentiating*.

In earlier work myself and colleagues have begun to trace the differentiation of rural space through the delineation of ideal typical conditions (see Marsden *et al.* 1993; Murdoch and Marsden 1994; Marsden 1995). As outlined in Chapter 5, variations in regional and local rural spaces are associated with the differential role and power of agricultural, residential and other commercial interests, giving at least four distinct types of rural social and political formations. We have termed these the: *preserved*, much of South-east England; *contested*, parts of rural Devon; *paternalistic*, parts of East Anglia and Lincolnshire/Northumberland; and *clientelistic*, parts of Wales and Cumbria, countrysides[1].

A key element of change in rural space is concerned with both the agricultural and *non-agricultural use and development of land*. Rural property rights, their exploitation, development and commoditisation have been shown to be an important locus around which social and economic change occurs (Marsden 1995). This occurs at different speeds and in different directions according to local and external combinations of relationships operating in rural localities. If one examines equivalent research findings and approaches in mainland Europe, for instance in Italy (see Iacoponi *et al.* 1995), the Netherlands (Van der Ploeg 1997), we can see that similar processes seem to be occurring even if the terminology (e.g. local systems, rural industrial districts, new farming styles) and regulatory systems differ. One element of convergence concerns the rearrangements of agricultural and non-agricultural property rights around new consumption and production dynamics, whether these are associated with the selective spread of ex-urban population and the concurrent demands these then place upon rural resources, or with new food consumption requirements. While considerable work has documented these processes in terms of farm pluriactivity (see Arkelton Trust 1993), scholars have yet to trace the multi-dimensionality of these processes of change either in terms of external dynamics or local response mechanisms. Similarly, while the endogenous development literature begins to wrestle with these issues (see Van der Ploeg and Van Dijk 1995), its focus upon the local as opposed to the interaction of the local with the non-local tends to produce an important but still partial analysis of the processes which are creating differentiation (see Lowe *et al.* 1995).

While then, there is a recognition amongst the main literatures about the integrative and holistic nature of the new processes of rural change, the approaches thus far have yet to theoretically develop an approach which begins to guide a clearer understanding of the processes which are *making things different* in the post-productivist countryside.

In order to overcome some of these limitations it is necessary to further explore the concepts of 'action at a distance' and 'action space' developed by Murdoch and Marsden (1995). We need to give more attention to the combination of local *and* non-local processes which impact *together* upon rural areas, and to assess how different local and non-local combinations configure in different rural spaces. In doing this we cannot assume that these processes will necessarily harmoniously interact to produce an integrated and functionally coherent rural space. As I shall argue below, rural holism and integration are by no means givens in the differentiated countryside. Indeed, we might argue that this is increasingly unlikely as different development processes collide in the same rural spaces, with different dominant social formations and regulatory authorities left to modulate and reshape these processes. In this way we have to conceive of rural spaces as ensembles of local and non-local connections, of combinations of local actions and actions 'at a distance', situated in regional economies and different institutional contexts. In this sense, different rural spaces have different combinations of networks to which they are connected. In developing an understanding of these we not only have to consider the lateral or spatial degree of differentiation (i.e. between the *preserved* and the *contested* bits of countryside). It is also necessary to combine this with an analysis to the 'vertical' chains of connection which are incorporated in each, through, for instance the largely 'at a distance' food chain relationships, or the wider markets of rural consumer demand associated with housing, tourism and leisure.

This begins to take us significantly beyond the dualistic assumptions of rural land development associated with either an agricultural or non-agricultural focus. It also opens the way to explore the inter-sectoral developments (for instance concerning the various agricultural land conversion markets) which have become increasingly relevant to the pace and direction of change in rural areas. What are the main parameters of these 'vertical' chains of connection in a British context? And, what light do they shed upon the variable regulation and particularly economic governance of rural space ?

Parameters in the new territorialisation of rural space

Taking our differentiated rural spaces (the *preserved, contested, clientelistic and paternalistic countrysides* outlined in Chapter 5) we can identify at least four key spheres associated with rural land development. These reflect different dynamics which are locally and externally generated, different production and consumption relationships, and different degrees and types of regulation. Their relative significance also varies according to the prevailing social and political constitution of the rural space (i.e. how it is aligned to the ideal types). We can thus begin to identify different internally and externally generated dynamics which have varying spheres of influence in the differentiated countrysides. The four key dimensions are: mass *food markets, quality food markets; agriculturally related development; rural restructuring (non agricultural development).*

An exploration of these dimensions allows a more specific focus upon the inter-sectoral constitution of rural spaces, and it raises some pertinent questions about what might be the most effective ways to regulate the differentiated countryside in the post-productionist context.

Mass food markets

Despite the uncertainties associated with the CAP, the growing health concerns of consumers and the potential effects of the liberalisation of global food trade, mass market food production still dominates the British rural land base (Marsh 1997). The majority of farmers are increasingly hooked into the vertically organised food chain which is dominated by the power of the corporate retailers and manufacturers. These sets of interactions and relationships are therefore national and global in character and are subject to technological changes (the most recent of note being genetically modified organisms) which demand intensity of production and scale economies. In addition, the possibilities of such innovations as genetic manipulation gives added influence to the downstream sectors in shaping the nature of land-based production and controlling production schedules at 'arms length' or; 'at a distance'. While less emphasis has been placed on this sphere over recent years by rural scholars, it is important to recognise its continued salience and resilience in the shaping of rural space.

Even where there have been concerted attempts to reduce this intensive model, such as through the MacSharry reforms of 1992, and the more recent proposals to reduce support prices and export subsidies (e.g. Agenda2000), the effects of the global orientation of these markets has served to give impetus for increased levels of intensification and concentration. For instance, in many parts of the productionist and *paternalist* countrysides of Yorkshire and Humberside and Lincolnshire, which largely lie outside environmentally designated areas, intensive production, farm concentration and new forms of lease-holding and contract farming have continued despite Macsharry. Moreover, any prospective liberalisation and dismantling of CAP is likely to reinforce this trend in many of the most agriculturally productive (and by definition globally competitive) areas. In addition, as Banks and Marsden (1997, 2000) illustrate with reference to the dairy sector, the abolition of former corporatist structures (such as the Milk Marketing Board, MMB) is tending to expose producers even more to the regulatory authority of the downstream manufacturers and retailers.

For the less agriculturally productive areas, that is in the *clientelistic* and poorer upland areas of Western Britain, the slow convergence to world market prices is likely to lead to the growing call for more subsidisation rather than less, for environmental and social welfare reasons, as much as for agriculturally productive ones. The regulation of mass food markets in rural Britain is likely to become increasingly bifurcated both in geographical terms and in terms of the articulation of rural interests as they relate to agriculture. Eastern lowland England may see long term advantages in the technologically engineered 'race to the bottom' global competitive scenario; but, for much of rural Britain – not least the areas of *preserved* as well as *clientelistic* rural regions – this is not an option they can win. The question will be to what extent can they resist and develop alternate agricultural strategies so as not to enter the race?

Quality food markets

For a variety of different reasons, and largely associated with changing consumer conditions, it is now clear that food markets are becoming more differentiated on the basis of a range of socially constructed food quality criteria. This is being reinforced by the intensive competition between corporate retailers who are developing hierarchical market positioning strategies which reflect socio-economic and locational characteristics of different groups of consumers (see Marsden *et al.* 1998). In addition, while the features of novel and locationally defined foods have been with us for some time. and are now very much a part of the rural development policy discourse, it is clear that in a more macro-sense the attribution of quality criteria along food supply chains is becoming much more ingrained.

This is most evident in the case of organically produced milk in Denmark (see Michelsen 1995), representing up to a third of all milk sales; and varied examples from Italy where organic food supply chains are beginning to take hold as consumers relearn their cuisines (e.g. Parmesan cheese, mozzarella etc.). It is interesting to pose, for instance, how the recent beef crisis across Europe is beginning to open opportunities for the organic beef sector. In Umbria, for instance, the virtual closing down of the mass beef markets following the beef scare has through necessity meant a growing emphasis upon developing more direct links between producers and butchers; short circuiting the more sophisticated but increasingly vulnerable retailer-based supply chains.

It is clear from such tendencies that the active involvement of near-market agencies in food supply (particularly the retailers) is going to further encourage the use of rural space for 'quality' food products, whether this comes from Sainsbury's supply managers enrolling specific farmers to produce contractually designed and 'free range' pig meat under welfare and extensive regulatory conditions (e.g. Sainsbury's 'tenderloin pork' produced on contract with producers in Humberside, reared in outdoor welfare friendly systems), or groups of locally inspired producers beginning to exploit new market niches on the basis that if it is locally defined it will sell to the careful consumer directly. There are now a plethora of farm and food assurance schemes being developed by: regional farmer groups themselves (for instance, the Scottish and Welsh Quality Assured Beef and Lamb schemes); by corporate retailers (such as Sainsbury's Partnership in livestock Scheme, or Tesco's producer group partnership); and by the representative organisations like the NFU, (the NFU-Retailer Partnership) or the RSPCA (Freedom Foods scheme).

The evidence begins to suggest that these tendencies are more significant than there mere novelty value or existing market share would suggest. Recent forecasts suggest a 20 per cent annual increase in the demand for organic products (Sainsbury's Supermarkets 1997) and severe problems of quality supply. And we have to ask why it is that despite the growing questioning of mass food production, such innovations have been so slow to take hold (particularly in the British case). Indeed, from a European perspective we can begin to see this developing as a major active force in market regulation and supply-chain innovation potentially linked to regional economic development benefits (Tovey 1997). One which is likely to provide new forms of comparative advantage for certain rural areas over others.

It is clear that the degree to which different forms of quality supply chain regulation is developing in Europe varies considerably. Moreover, this variability can be seen as the emergence of different regionally induced 'insurance policies' against the more aggregated mass food 'race to the bottom' scenarios portrayed above. Indeed, one evolving story of European food policy could be the gradual development of regional quality parameters which aim to protect local farming systems and project regionally identifiable and quality assured foods to the more sensitive consumer. One notable example of this variability concerns the national use of the EU 1992 Designation of Protected Origin (DPO) regulations. These promote opportunities for local supply chains in the European internal market. The principle is to distinguish products which are derived from methods of production and specific geographical origins. For instance, Italy held over thirty cheeses with the certificate and nearly a hundred products overall by 1996.

These new systems of food supply and production require revised organisational and innovative forms of partnership in different rural localities. It follows that the regional and local capabilities and *capacities* to organise such partnerships may be much stronger where there are longer traditions of co-operation and collaboration. In much of rural Britain the combination of a dominant intensive model of supply regulation through the operation of the CAP, together with the growing power of national and international retailer-led distribution systems, has tended to constrain the opportunities for local initiatives and supply chain innovations. Nevertheless, for those *clientelistic* regions which have traditionally been rendered agriculturally marginal (e.g. Wales and parts of Cumbria), the construction of such quality markets could become significant insurance policies during a period of uncertainty over the longevity of existing upland farming support structures.

In a period of 'careful consumption' and uncertainty in mass food markets, rural areas are likely to be differentially incorporated into the *quality food markets*. These will be much more dependent upon the attribution *by others* of standards and conventions of food production and transfer. This will further integrate the traditional producer groups into compromises with consumer demands and lead to new influences upon the uses of rural land and territorial differentiation.

Agriculturally related changes

Probably most effort has been placed by researchers in specifying the nature of agriculturally related changes in land development in rural areas over recent years. This is generally defined as 'agricultural diversification' and tends to concentrate on the new activities farmers participate in, including converting and developing land and buildings. However, this process also requires the development of new connections and networks. The actual agricultural diversification outcomes and their degrees of success, are based upon new local, regional and national relationships being forged between the farm and these wider networks.

It is clear that these developments are highly locationally specific and that they relate to the particular growth in non-local and non-agricultural markets. In many ways it tends to increasingly highlight the authenticity of the rural through the refashioning of former 'dusty' rural artefacts, while at the same time exposing these to a completely new audience and marketised demand. It is thus highly linked to the vagaries of the urban wealthy.In the *preserved* countrysides of the South east, the recent bonuses received again by successful city bankers and dealers are already creating another round of commoditised rural development for exclusive rural homes, many of which are converted farm houses and buildings.

These rounds of rural fetishism, over the long term lead to, in absolute terms, the continued urbanised construction and commoditisation of the rural. This process of reconstruction takes highly differentiated forms both within and between nation states and rural regions. What a prematurely retired accountant or banker may wish to do with his surplus capital may differ considerably between nation states and between regions. The rural arena for these groups continues to represent a world of highly positional new choices.

More generally, however, because of the historically large amounts of capital, property and potential consumption resources tied up in agricultural land, it is inevitable that a major intersectoral arena around which rural change and differentiation will continue to occur concerns land developments which transform former agricultural assets. Agricultural property becomes a differentiated resource pool which both residential and tourist consumers can exploit. Evidence would suggest that in the British case, we are still at a fairly early stage of development. For instance, despite the growth in farm diversification and the considerable policy and academic emphasis it has received, we should remember that only 75,000 hectares, or 6.6 per cent of the total land area of the British countryside is available for recreation by legal right, for instance as country parks, national parks common land etc. In addition, there is linear access provided by 200,000 KM of public rights of way. So far, only 15 per cent of farmers are engaged in the provision of tourist enterprises, accommodation and holidays; although this rises to 23 per cent in the *contested* countryside of the South West of England, and 35 per cent for the county of Cornwall.

Although the number of non-residential visitors has grown relatively little since the late 1970s, the amount spent and the range of activities pursued has increased significantly. There is an expanding spectrum of novel activities associated with informal recreation and commercial facilities. These range from the development of theme parks, holiday villages, golfing complexes, holiday cabins and war games. There are over 400 sites for the latter activity, with rural landowners earning up to 15000 pounds for a good facility.

These trends are set to continue, and given a potentially more compliant planning system outside the *preserved* and *contested* countrysides, this could lead to considerable development out of agriculture and traditional rural activities. In the *clientelistic* rural regions this is being seen by government and rural development agencies, local planners and promotional groups as a main panacea to the problems of rural deprivation and jobs. In these areas, tourist and recreational facilities on converted agricultural properties represent substantial proportions of investment in the European OBJECTIVE 5B funding for rural areas and in LEADER and other local authority programmes. In the Midlands Peak District and in Cumbria, local initiatives are attempting to develop employment opportunities around the forecasted increase in tourist demand. In Wales, the Wales Tourist Boards'(WTB) 2000 strategy (1996) is promoting holiday tourism as a way of boosting jobs and on-farm diversification. 'The tourist' represents a larger proportion of rural GDP than agriculture, and in collaboration with Leader groups the WTB are aiming to promote the 'quality product- Welsh country holidays' and to attract the foreign visitor.

In *clientelistic* regions like Wales, with a traditionally marginal agriculture, fragile rural economy and a plethora of agencies attempting to obtain diminishing sources of state aid on an intensely competitive basis, 'selling the rural environment of Wales' is increasingly seen as its greatest asset. This means that tourism and recreation promotion becomes a major element in rural economic governance, and that this is predicated upon the continued opening up of former agricultural space and resources to external groups and a wider range of commodified services. At the same time, however, the same rural environment has to be maintained as a quasi-agricultural space. This agricultural conversion process is market led and agency regulated. Its priority is to continue tourism related employment (from 7.7 per cent to 11.1 per cent between 1981-91) and encouraging tighter investment gains concerning the regional and local investment of state funding. Since 1985, the WTB has spent 32.6. Million in assistance, creating 4300 jobs.

The growth in agriculturally related land development, associated with the meeting of new residential and tourist demands is increasingly seen as a major panacea for the rural and regional development agencies in the *clientelistic* countrysides of Wales and Cumbria. As agricultural employment falls continuously, developing rural tourism through the necessary redevelopment of former agricultural assets becomes a major rural development goal.

Rural restructuring (non-agricultural)

Over many rural spaces it is clear that the main forms of restructuring of the land base has little to do with the agricultural sphere in a positive sense. Rather, they are associated with the extra-agricultural processes which are attracted to the exploitation of the redefined rural resource. The exploitation of 'nasty' developments such as mineral working, landfill, and toxic waste, as well as defence establishments, means that rural land rights and relations are somewhat sporadically vulnerable to externalised investments and development which attempts to exploit the physical and social resources of rural space. These are, in the *clientelistic* countrysides at least, often given conferment by local and regional economic development and planning agencies which wish to increase employment potential for the local population. Here, unlike in the *preserved* and *contested* rural spaces, local planning is more pliable to inward forms of investment and development. In West Cumbria, for example, the opportunities to attract open-cast mining, toxic waste sites, industrial processing and large-scale tourist developments means that implementing environmental conservation policy is all the more difficult. Moreover, with the absence of an agricultural justification for restricting industrial development, there are greater opportunities for the selective siting of industrial and retail parks. With the absence of a large preservationist residential middle class (as in the *preserved* countryside) conservation planning has to rely upon environmental justifications associated with the need to attract more tourists to the region. These justifications can seem weak when job creation possibilities from industrial developments are at stake.

In the *preserved* countrysides, however, with their greater potential for organic as well as inward economic development, the planning system becomes a major mechanism for restricting such development in the open countryside and in most of the villages and small towns. Here, rural space, as Murdoch and Abram (1997) indicate, becomes middle-class residential space. Where small-scale industrial and service developments do occur (see Murdoch and Marsden 1994) they are privately-led by entrepreneurs and a collection of private exchange agents, consultants and advisors.

By contrast, in the *clientelistic* areas, rural restructuring is heavily publicly underwritten and structured. Emblematic in this regard is rural Wales, with its plethora of development agencies including the WDA, the Development Board for Rural Wales, local TECs and a series of more local agencies (such as Menter Powys, Antur Tiefi and Antur Cwm Taf). These have generally prioritised inward investment through providing attractive publicly funded packages for incoming manufacturing industries. They have been able to create new 'hot spots' (such as the Newtown-Welshpool corridor) in traditionally agricultural areas. While this has led to a growth in manufacturing employment, weekly wage earnings have been lower than in urban areas and they have not sufficiently shifted the rural dependence off publicly funded employment. In rural Wales and in West Cumbria, despite the past twenty years of government restraint and reorganisation of public expenditure, these areas are highly dependent upon the increasingly selective release of public funds for sustaining their fragile labour markets. In rural Wales, the public sector is still the largest employer (28 per cent). Widening gaps develop between 'hot' and 'cold' spots within the same rural regions (for instance, between east and south east Wales and the South West, or between South-east and West Cumbria). This means that the 'colder' areas are still highly reliant upon both a declining publicly funded infrastructure and an agricultural sector, as discussed above, which is facing growing uncertainties.

Rural restructuring processes are therefore highly influenced both in their intensity and in type by the varying institutional and regulatory structures and processes developed in the different types of countryside.

This is, most clearly demonstrated when comparing the *preserved* and the *clientelistic* countrysides.

Regulating and governing the differentiated countrysides

Obviously, the four dynamic spheres of development outlined here are generalised attempts at identifying some of the key economic processes impacting upon the differentiated countrysides in the British and the wider European context. Within each of the them it will be necessary to specify the more specific strands of commodity and network relationships which contribute to the overall sphere. This will become an important task in developing comparative rural analyses of the differentiating countryside.

Nevertheless, their broad delineation helps to bring into sharper focus the degree to which different proportions of each sphere become significant in shaping spatial rural differentiation more generally. In the *paternalistic* rural regions still largely dominated by intensive agricultural production (much of Eastern England), it will be clear that the *momentum* of change in the *mass food markets* and to a lesser extent the *quality food markets* will be central to the further restructuring of these places. *Agriculturally related and rural (non-agricultural) restructuring* may also be occurring but it is as yet peripheral to the trajectory of differential rural development experienced.

By contrast, the regions of *preserved* and *contested* countryside now owe their development *momentum* to the *non-agricultural development* forces of *rural restructuring*. And the question for many of the local residents becomes one of spatial resistance - through, for instance the use of and participation in the planning system and formal preservation societies – to potential forms of development. Ward *et al.* (1995) and Flynn and Lowe (1994) describe vividly, for instance, how in the *contested* countryside of rural Devon, agricultural concerns are increasingly voiced by the newer rural non-farming residents as more of them adopt a preservationist and conservationist position, involving the newer environmental agencies like the National River Authority (now the Environmental Agency).

The different and relative combinations of the four spheres in the differentiated types of countryside act as agents for the further generation and *momentum* of social and economic changes in these spaces. In the *clientelistic* countrysides, *quality food markets* and the *agricultural and non-agricultural restructuring* processes are jostling for dominance at the more micro spatial levels, and it is in these countrysides which are less certain of how these different spheres will relate to each other, or indeed, as to which will become more dominant in the longer term.

The four spheres identified also exhibit different production, exchange and reproduction features such that some may be at different phases of development at any one time. Rural spaces are thus constituted with these four spheres, each holding a relative position to each other and having their own temporal and regulatory dynamics, involving different networks of actors, agencies and relationships with local, national and global markets. We need both for analytical and policy reasons to move away from the strictly geographically defined notion of 'local rural area' in this sense. Rather, we need to conceive of differentiating rural spaces which are caught up in different webs of local, regional, national and international supply chains, networks and regulatory dynamics. For instance, in the *paternalistic* rural areas where mass agricultural markets are predominant, they will obviously be participating in a global set of terms of trade, of networks and regulatory dynamics, compared with the *preserved* or *clientelistic* countrysides.

In comparative terms we need to be able to identify these spheres and their relative positions in different rural spaces, assess how each of them lead to aspects of change and status in rural space, and the degree of complementarity or conflict between the spheres. So far we have insufficient data gathered on these vertical and lateral networks that impinge upon and run through rural space, partly because of the traditional sectoral orientation of our survey methods and public accounting systems. For instance, the current batch of Rural White Papers may celebrate rural diversity, and the merits of a holistic approach, but official definitions of variation still cling to physical distance and centre-periphery type classifications of rural space. The analytical framework explored here, therefore, carries a considerable empirical burden if we are to fully understand the new patterns of rural differentiation (as a series of regionally contextualised ruralities) and the role of social actors and institutions in it.

Nevertheless, we can begin to map out some of the key features for the differentiating countryside, and to identify particular areas which represent increased uncertainties for different rural spaces. In addition, some important questions are raised concerning the relative empowerment of different social actors (such as ex-urban middle-class groups in rural Buckingham (see Abram *et al.* 1996), or food producers in East Anglia (Newby *et al.* 1978). Within each of the differentiated rural spaces different local/non-local social configurations of networks and actors are developed, and these are aligned to the separate development spheres identified here. These configurations allow relative power to be distributed in different ways, such that the 'power geometry' of each rural space creates different governance and regulatory issues and processes. For instance, in the *paternalistic* and agriculturally dominated areas of Eastern England, farmers and landowners are increasingly having to accommodate the private-interest models of regulation and governance which is led by the corporate retailers. These private-interest power relationships are in direct contrast to the local/non-local configurations in the *preserved* and the *contested* countrysides. Here, local residential interests find empowerment with their articulation with local planning; which is, of course, supposedly a public-interest system of regulation. Hence, the rural governance and power geometries of the different countrysides are associated with the development of these local-non-local configurations, and their alignment to the four spheres of rural development outlined here.

This would suggest that there is an interactive set of relationships between the production and reproduction of different rural-based social formations and the configuration of the development spheres in the differentiated countryside. Moreover, it also follows that each rural space will have a different composite set of relationships with consumers of rural resources through their links with the different markets and regulatory institutions.

It should be quite clear from this discussion that to understand the nature of contemporary rural differentiation it is necessary to refine our conceptual parameters in such a way that we do not reduce interpretations to singular categorisations based either on externally generated structural concerns or aspects of local empowerment alone. The emphasis needs to be placed on integrating local and external processes and assessing how they are expressed in different regional ruralities. Two major dimensions which are important in this process relate to (i) the nature of supply chain regulation and their dynamics and (ii) the degree and character of institutions and regulation associated with the different spheres of development outlined above. For instance, there are in general terms considerable variations in the spatial and social regulation of these spheres. *Mass food markets* are largely internationally and corporately regulated and are likely to be more so as trade liberalisation reduces support measures and differences between EU and world market prices. On the other hand, *quality food markets* raise the spectre of local organisation and the potential development of regionally embedded supply chains and co-operative organisation. Much of the *agriculturally related restructuring* has landed, in some cases (as with golf courses) with something of a surprise, in the form of development proposals on the desks of the local planners office. The land-use planning system thus becomes a major regulatory arena (see Lowe *et al.* 1993) for the allocation of agriculturally related developments. These differences in the type and origin of regulation in the development spheres holds important implications for the broader and more sensitive understanding of both the private and public governance of differentiated rural space.

The development spheres highlighted here are not simply economic drivers. They bring with them an ensemble of regulatory instruments, processes, practices and demands. These have to be organised at different spatial scales. This means that the governance of rural areas is inevitably uneven in its composition within each type of countryside and more explicitly between the differentiated countrysides. Moreover, we have to also realise that an interpretation of the basic structural and institutional characteristics of such variable rural governance is not sufficient in uncovering the processes of uneven development. It also needs to accommodate the processes and practices of regulation and governance within institutional structures. For instance, the same types of planning frameworks can have different results in the different countrysides. Similarly, economic development agencies' organisation can be replicated but the results can differ markedly. It is the *substance* of governance – the social regulation of practices and processes which run through institutional structures of rural governance – which becomes an important focus in the remoulding of rural spaces. Moreover, it is the *blend* of the different and not necessarily complementary regulatory processes and practices, associated as they are with the main development spheres identified above, which begin to shape regional ruralities.

Within the differentiated types of countryside identified we can see, therefore, that different internal configurations associated with at least four different land development spheres are helping to shape the process of uneven development of agrarian space. The question of regulation – the power of authorities to act – is therefore related to the dynamics of supply chains and the process of institution building in and beyond rural spaces. What is increasingly noticeable about these dimensions – despite growing rhetoric about enhancing local participation and power bases – is the 'at a distance' influence of external actors and their effects upon creating demands for rural resources. Rural development is increasingly reliant upon demand conditions created outside its geographical boundaries.

Conclusion: differentiation, integration and ecological modernisation

Over recent years, there has developed a considerable consensus in the policy and the academic debates on rural policy to progress a more 'holistic and integrated approach' to rural areas and to the regional economies and spaces in which they are increasingly embedded. This is reinforced by the best examples and experiences of recent 'bottom up' and partnership rural policies such as LEADER and Objective 5b, and the growing requirement to move away from the traditional sectoral approaches dominated traditionally by agriculture. In addition, after the untrammelled growth – partly as an effect of the breakdown of the hegemony of this sectoral approach – of agencies, quangos and government offices as well as local authority – there is (as we saw in the specific case of Wales in Chapter 6) a growing realisation that organisational integration is as important in dealing with rural needs as the development of integrated policy initiatives and actions themselves.

contain its environmentally destructive features at the same time as stimulating its potential to meet diversified consumer needs.

The post-productivist paradigm does not, therefore attempt to radically solve the problem of agri-industrialism. Instead it acts very much as a palliative paradigm, by containing it. It does not address the question of a different sustainable development pathway for agriculture. Rather, it suggests consumer-led ways around its pathological features. By the end of the 1990's these features were becoming more evident in both research and policy terms. There was a growing realisation that post-productionism, by itself was not the saviour for agriculture , or more particularly for more sustainable food supply and consumption. However small the agricultural base became in employment or GDP terms, it could not be left as the residual appendix of the rural economy for a host of social, economic and ecological reasons.

Section 3 of the book addresses these questions directly by reproblematising agrarian nature. It outlines and explores the emergence of a new and more tentatitve rural development dynamic which begins to re-centralise agriculture in new ruralities and experiences. Its main distinguishing feature is that it focuses analytically upon agriculture in more diversified rural economies (as with post-productionism), but it does so by also centralising the potential value of agro-ecological practices as part of the new condition of rural sustainability. In this sense, it attempts to 'think the unthinkable' by exploring the marginal; that is how can regulatory frameworks and practices be put in place which potentially reverse the tendencies outlined above. The following chapters begin to chart some of the theoretical and empirical components necessary for developing such a paradigm.

Chapter 8 begins by attempting to position some theoretical tenets of social and agro-ecology in the context of wider debates concerning ecological modernisation theory. Chapter 9 and 10 then empirically explore these issues with reference to research in the UK. Chapter 9 examines some of the redefinitions of the farm business which are needed to progress such a paradigm. Chapters 10 and 11 then look more closely at the nature of alternative food supply chains (short-supply chains and organics) which are developing in the European context; and particularly assess these in terms of their likely trajectories with regard to some of the principles of agro-ecology. In these ways these explorations at least begin to shape a rural development paradigm which re-centralises agriculture into a new (ecological) modernisation trajectory both for rural studies and for rural (and urban) society. These new symmetries are addressed in the conclusions. A revamped rural development paradigm is one option in developing more rural sustainable conditions and in progressing the 'condition' of rural sustainability. It will however, have to cope, quite competitively with the more dominant agro-industrial and the post-productivist logics outlined and explored earlier.

8 Redefining agro-food and rural development: some explorations into agro-ecology

Introduction: EM or GM?

The study of agro-food has now become a popular sector for the exploration of both political economy and social constructionist perspectives. The latter tend to argue that the former are content with a narrow, binary and modernist perspective on nature, and have restricted its social and political significance in late modern or post-modern societies. The agro-food sector (and its links to rural development) has been staging an important battleground between these competing perspectives.

In this chapter I want to suggest something of a re-definition of political economy of agro-food, coming after, and in full acknowledgement of, the contribution that social constructivist approaches to nature have made.

I will argue that significant shifts and contestations recently characterising the agro-food sectors of advanced countries (at least) suggest the need for a revised political economy which takes forward a more agro-ecological and socio-spatial approach. It should do this by building upon the major strengths and weaknesses of social constructionism and challenge more vigorously the assumptions lying behind many of the existing and dominant 'modes of thought' currently influencing the direction and dynamics of agro-food developments. These dominant and social science-based modes of thought have created, in themselves, significant barriers for progressing sustainable agro-food development.

At a slightly lower level of analysis the chapter will address the inadequacies of ecological modernisation approaches to agro-food with reference to the *agro-industrial dynamic*. What does EM have to tell us about GM? Can the agro-industrial complex ever achieve higher levels of economic development with lower (more sustainable) levels of environmental impact? Or, as Mol (1999) suggests, lead to a 'reflexive institutional reorganisation of industrial society which attempts to overcome the ecological crisis'?

I will then identify and outline two other modes of thought and perspective which are relevant for positioning agro-food; what I call the *hygiene/bureaucratic mode* and the *social constructivist mode*. I will position the specific definition of nature within these modes, demonstrating that they too tend to deny holistic and material conceptions of rural and agro-food nature which can be progressed in a sustainable fashion. Indeed, as I will argue, the emergence of these modes of thought have tended to legitimise and empower the domination of the *agro-industrial mode* of development and its vast array of scientistic techniques. In conclusion, I begin to outline a revised political economy of agro-food which adopts, adapts and progresses some of the key agro-ecological principles. This requires the need for detailed theoretically driven, empirical research on the ways in which actors and institutions in agro-food supply chains are reconstructing social nature by assembling new social and natural resources and rights (see Van der Ploeg *et al.* 2000).

Ecological modernisation: what does it contribute in the agro-food context?

The onset of GM technologies over the past few years have been portrayed by their proponents very much as a realisation of the benefits of ecological modernisation. Most variants of EM theory tend to claim that, in contrast to earlier neo-Marxist analyses, environmental conflicts can no longer be interpreted as following predictable paths with static opposing parties and interests, or class and distributional effects; but, it is argued, they can be resolved, in part, by industrial modernisation, rationalised through its armoury of applications of experimental science. As Buttel (1997) depicts, it provides a framework which assumes the modernisation of capitalism beyond the conceptions of the treadmill of production and 'the growth machine'. To some degree, therefore, can we not see GM as synonymous with the more neo-classical variant of EM?

There are, however, a considerable number of different shades and guises that the perspective of ecological modernisation adopts which make it difficult to assess in relation to the agro-food sector. This extends from the neo-classical, Porter hypothesis, which sees employment and competition advantages from the growth in the use of clean technologies and environmental regulation, to the more sober treatment of EM as needing to move far beyond a simple conception of industrial production rationalisation to fully embrace the 'precautionary principle' and an enhance role of the state (see Andersen and Massa (2000). These authors argue, for instance:

> *'The acknowledgement that an eco-modernist rationalisation is unlikely to come about on its own, but will follow only if nurtured through a deliberate and far-sighted intervention policy, seems to suggest the need to confront some of the neo-liberal rhetoric inherent in the ecological modernisation literature. 'Win-win solutions' and 'getting the prices right' do not follow on from laissez-faire, but require intervention of a scope and quality that is likely to*

strain the capacity of many a regulatory state. In our view, the EM debate should, therefore, end the subscription to efficiency rhetoric, because it needs to begin to address issues of governance capacity much more explicitly, consciously and cautiously' (p. 344).

I want to return to this more embellished conception of EM towards the end of the chapter, after a fuller analysis of the agro-food sector, and the particular theories of nature that it implies. I argue, that in relation to agro-food developments in the advanced world, conceptions of *ecological modernisation need serious scrutiny and revision* given that they can be used in different ways to justify different types of agro-industrial development, including GM. Alternatively, I will argue that *real EM* will only be possible if we adopt agro-ecological type principles which do much more than simply tinker with the agro-industrial process. *Real ecological modernisation in the agro-food sector, I argue, means re-embedding physical and social nature into rural development processes*, rather than simply adapting production and consumption around a 'more environmental' set of adjustments. I will show, with reference to the 'post-BSE' and 'post GM' agro-food landscape, that attempts to develop *partial ecological modernising principles and policies tend only to reinforce an unsustainable form of agro-food development*. In this sense, certain EM solutions are not solutions at all. In the agro-food sphere we should be reminded that social-natural elements are embedded both in the means and conditions of production and consumption – from 'conception – to digestion' – and as such, more profound EM principles need to be constructed which do not compartmentalise nature into one of a series of specific negative outcomes to 'be solved'. In short, neither factory-bound, laboratory–bounded, or, indeed professionally-bound 'solutions' are adequate over the long-term.

Agro-industrialism and the panacea/ palliative of GM

As Barbara Adam has argued:

'*GM of food is understood as the crowning glory of a long history of time rationalisation… Taylorism to rationalise individual workers' actions; the 'just-in time system' to cut from production all elements of non-productive which means non-profitable time (and space); geno-technology to eliminate from the breeding process the time of generational succession…the genetic engineering of food is about money. It is about the promise of massive time saving in the scientific production of change. This is achieved by controlling time: controlling maturing, ripening, ageing and decaying; controlling the seasonality of animals and plants, controlling generational sequencing and reproduction. Control of time in the production process is an integral part of the success of industrial capitalism and so is the control of nature and the geno-technological control of the processes of life. In a context where time is money, geno-technology holds the process of very big money indeed. It has the potential*

> *of realising the time rationaliser's dream: instantaneous change in unlimited quantities, effected not at the phenotype but the genotype. At a stroke, changes introduced in the present alter the life-course and evolution for ever. However, this is also where we have to locate citizen and consumer unease with GM foods'* (Adam 1999, p. 139).

And as Graham more soberly reminds us:

> *'Genetically modified foods have already reached the European market. Most of us have probably eaten them, or foods containing genetically modified ingredients, even if we are made aware of it, and have not specifically chosen to do so'* (Graham 2000).

The controversies surrounding the onset of genetically modified organisms in the food system has tended to focus upon their potentially revolutionary impact in replacing non-modified crops and livestock, their seemingly uncontrollable spread across agricultural landscapes, and the potential effects upon flora and fauna. These concerns are controversial and are likely to run for a long time. While growing demands for more regulation of GMOs in the food supply system are likely to lead to restrictions and delays in the use of the technologies (partcularly in Europe), it now seems clear that they will remain, in one form or another, as a key aspect of agro-food innovation for the foreseeable future. The acreage of GM crops has expanded 25 fold between 1996 and 2001; and despite consumer and environmental unease, the industry has convinced almost all governments and world bodies to support variable packages of these technological innovations. I want to concentrate here upon what the dawn of GMO foods and technologies are likely to mean for the broader agricultural and food sector. As a recent report has confirmed:

> *'The rate of adoption since 1996 of commercial GM crops in the three principal growing countries, especially the US, is remarkable. From a standing start in 1996, glabally an area twice the size of the United Kingdom's landmass is under commercial GM cultivation. During the six-year period 1996 to 2001, the global area of transgenic crops increased more than 30-fold, from 1.7 million hectares, in 1996 to 52.6 million hectares in 2001. GM crop cultivation is still increasing, but not as fast as at first. In 1999, there were 39.9 million hectares of land around the world under cultivation with GM crops. This represents an increase of 44 per cent or 12.1 million hectares compared with 1998, when 27.8 million heactares were grown. The rate of increase between 1999 and 2000 slowed. In 2000, 44.2 millian hectares were sown compared with 39.9 million hectares in 1999, an annual increase of 11 per cent compared with 44 per cent the previous year. The area rose 19 per cent again however between 2000 and 2001, to 52.6m hectares, almost twice the corresponding increase between 2000 and 2001'* (AEBC 2002, p. 25).

Despite the uniqueness of the GMO technology in scientific terms, both in technique and in its potentially generic impact upon food producers, suppliers and consumers, in general terms, they provide the next innovative basis for a further stimulus for the development of the 'industrial model' of food production and supply. By this I mean the dominant food supply chains which produce standardised products on a capital intensive basis, giving emphasis to the optimum (quantitative) level of production. Such a model in the North has been developed with significant amounts of public funding during the post-war period, but supply is now largely controlled by large multinational food corporations and retailers.

Along with other biotechnologies, GE provides a new terrain for the progressive reduction in costs in these industrialised food supply chains which then creates opportunities for further 'economies of scale' in food enterprises. In other words, the 'technological treadmill' on which producers and many smaller manufacturers are placed takes another turn; leading to the further concentration of production amongst smaller groups of large-scale producers and manufacturers. It is likely that, for instance, the larger, more intensive farmers are going to be the first to purchase the new GE seed varieties and it is these who will gain from increases in production and the subsequent reductions in the other input costs which are supposed to arise. Plant breeding and seed reproduction will become further removed from producers with individual firms tying groups of intensive producers to the annual purchase of their GE modified seed varieties. Those who resist adopting these varieties will face increased costs and be either forced to adapt or go out of business.

This process is far from new in post-war agriculture and it has been a major reason why we have seen the radical growth of large intensive farms and the progressive disappearance of the smaller, mixed farm. GM technologies are thus likely, particularly in the context of the World Trade Organisation (see Gibbs 2001), to further 'speed up' the structural change in agriculture and food supply, making it more difficult for smaller producers to stay on the land (Lappe and Bailey 1999).

Much of political economy perspective on agriculture and food over recent years, however, has challenged this technological/ industrial imperative (see Buttel 2001; Goodman 2001), but it is now time to develop more rigorous normative frameworks for developing the much needed alternative and 'non-conventional' sustainable food supply chains based upon all of the three spheres of sustainable development (environmental, social and economic), and incorporating the public-interest-consumer concerns. Growing food consciousness and food politics (arising not least from the handling of the GM controversy) will begin to assert different definitions and equations concerning the relationships with nature, region and food quality. '*Economies of scope*' (Van der Ploeg *et al.* 2000) rather than of *scale* alone could become more important in this model of food development, whereby producers and rural regions offer pluri-active /agricultural and rural development products and services, rather than continuing with monocultural agricultures. The evidence is mounting now that smaller farmers who can add value to products and manage alternative and multifunctional enterprises on their land can make a profit over the medium and long-term. The onset of GMs raises the need to urgently develop better and more practical ways forward for producers in achieving these new 'economies of scope'. This represents a more *'orchestral'* rather than *mono-functional approach* to the countryside but to really develop it will need more direct support from governments. Rhetorical statements concerning the need to develop sustainable agricultural practices while also opening the door to genetic technologies will no longer do. A more coherent and strategic rural/food and agricultural policy will be needed which does not avoid, but faces these current contradictions in policy development.

Food governance, entitlements and the limits of current state policy

Given the demise of the post-war food policy which was based upon a strong national government steer in both the enhanced production of foods for the mass population, and the establishment of public policies to ensure food safety baselines, efficient intake and nutritional standards, current governments face a key dilemma which is heightened by the GM debate. This concerns the following conundrums concerning the effective balances between state intervention, corporate interests (particularly food retailers and manufacturers), and the now more diverse public consumer interest. Over the past two decades the State has progressively handed over responsibility for key sections of the regulation of food quality to the corporate food manufacturers and retailers. A private-interest model of food regulation has developed alongside the public local authority enforcement system, such that retailers are now much more effective in controlling the quality and direction of foods from 'plough-to-plate.' The rise of private-interest regulation does not, however, avoid the need for the State to protect the consumer interest (as the BSE and the GM controversy shows) (see Marsden *et al.* 2000b). What it does do is to limit the state's ability to act in the consumer and public interest. Moreover, the progressive privatisation of research and development associated with food (expressed, for instance, in the proliferation of private patenting of genetic modifying techniques) means that governments can no longer 'command and control' the food system on behalf of consumers.

The need, given these conditions is for the State, corporate interests and consuming interests to develop new sets of relationships which build upon the different knowledges used and applied by each. This is no easy task; and the reorganisation of public food regulation through the setting up of the proposed Food Standards Agency (in the UK for instance) will only be partial in this regard.

GM technologies in foods may increase the quantity of foods available, and in some cases might improve the quality in relative terms. Its 'ecologically modernising' features suggest, for instance, at least for possible environmental and public benefits: (i) crops which are protected against insect damage and reduce pesticide use; (ii) herbicide tolerance, allowing farmers to control weeds without harm to the crop; (iii) disease resistance, by 'arming' plants against the destructive viral plant diseases with a 'vaccine'; and (iv) in a more long-term sense, nutritionally enhancing food which could give higher levels of nutrients, vitamins, and other health releasing properties. However, state agencies and consumer interests (both representing the public interest in foods) need to assert a more powerful role in considering the allocative and entitlement implications of such technologies. GM will not reduce food poverty, food scarcity and entitlement questions. Indeed, I would expect the uneven development of food production and consumption to intensify both between the North and the South and with each type of region. Moreover, it will reinforce a retailer-led form of food governance in Europe.

These processes of uneven development in the production and consumption of foods are, however, also likely to encourage the development of alternative food politics and social movements which will attempt to counter the private-regulated system of retailer-led food governance which has developed, as well as the broader industrial model of food supply. During much of the post-war period, the combination of state and corporate interests in foods have been successful in publically 'de-politicising' food. There are now clear signs that this has now come to an end, and we are facing a new more turbulent period of food governance. This new period seems to have at least two major dynamics: the intense and continued industrialisation of foods through the advances in GMs and bio-tech, and second, the growth of alternative systems of food provision which are currently far less organised or co-ordinated. Government and corporate retailers are two of the key agencies in managing these dynamics in the public interest. The controversial arrival of GM foods will continue to politicise the food agenda and it will force government agencies to take a bigger stake in the food system as it did after the Second World War. This time, however, it will not only be the advancement of an industrial model of food that will be on the policy agenda. It will have to examine and act upon the gradual realisation that foods are socially and economically manipulated, but that they are also natural products after all, just like the humans that consume them. A more socio-natural (sustainable) approach to food – one which places people, places and nature as central – could begin to make severe 'dents' in the industrial, standardised model; but not unless governments take a more interventionist and strategic role. The character and use of all technologies is determined not so much by the technology itself as by the social, political and institutional context in which it is placed. The arrival of GM places another serious question of the trajectory of the industrial food system and the degree to which governments are committed to real sustainable development in the agro-food sector.

It is important to recognise that, the modes of thought developed by social scientists, associated with agro-food, have an important role to play in these sets of contestations, just as they did during the development of the New Deal in the US and the diffusion of the industrially-based and state supported productivist regime in the European post-war period. There are three modes of thought which are currently particularly pertinent in demonstrating the obstacles that now confront the real development of an alternative ecologically based agro-food system. The main concern here is to locate the specific conception of nature in these dynamics and to recognise how these conceptions provide obstacles towards achieving more sustainable rural development objectives. In short, as Drummond and Marsden (1999) conceive, these modes of thought play a crucial role in 'maintaining the unsustainable' in both an economic and regulatory sense. Hence, it is not so much that the current dominant 'unsustainable conditions are not, in themselves self-destructive. Rather, it is the case that theses conditions constantly require sets of regulatory and scientific conditions to uphold their contradictory social effects. The foregoing modes of thought are therefore represented here as significant obstacles to the pursuance of agro-food and rural sustainability in that, in their different ways they have tended to 'uphold the unsustainable'.

Obstacle 1, social science in reverse: the agro-industrial model – the appropriation and subsumption of nature

As I explored in Section 1 of this book, the aforementioned agro-industrial model of food production is upheld and reinforced by an alliance of agricultural economists and biological scientists who are still legitimating the intensification and scale economies imported from American and Australasian modernisation theory. This sets a precedent upon reducing the costs and prices of primary products through continued technological adoption, corresponding reductions in production costs, and continued scale enlargement to reach economies of scale. Large areas of rural space are unable to compete in this 'race to the bottom' scenario and so are forced to go to the state for more and more support which, in turn acts only to moderate the effects of this treadmill. Under these conditions rural and agricultural nature is seen as an obstacle to be overcome; something to transform along industrial lines and logics.

The agro-industrial mode is strongly associated with the continued efforts of producers and manufacturers to reduce the importance of nature in the food production process. The disconnection of food production from nature is mediated by two interrelated processes: first appropriation, that is the attempt to replace previously natural production processes by industrial activities; and second, substitution, or the attempt of industrial capitals to replace natural products in the food system with industrially produced substitutes (Murdoch *et al.* 2000).

Increasingly this agro-industrial model is caught up in a bigger globalisation picture associated with the development of the biosciences research and industries (see Gibbs 2001). As Gibbs argues:

> '*the Biosciences sector illustrates the complexity involved in the processes of globalisation and the interaction between spatial scales of action. The example of GM foods and seeds provides evidence of contested notions of 'progress', 'nature' and 'safety'. Much of the power lies with the large transnational bioscience companies, enacted through trade agreements, while the evidence is that local and national resistance have only limited impact*' (p. 255).

Obstacle 2: the bureaucratic 'hygienic' mode – accommodating the consumer-countryside

While the agro-industrial logic is still in a fairly dominant position in global agricultural terms (e.g. as represented in the WTO talks), it has been supplanted in some respects by a parallel more bureaucratic mode of regulation and perspective. This has, in highly sophisticated ways, attempted to 'correct' the agro-industrial model rather than construct viable alternatives to it. It has done this by accommodating and assuming much of the theory of Beck and Giddens' notions of the 'risk society' (Beck 1992), by attempting to put into place, in a highly interventionary and bureaucratic fashion, policies which effectively 'police' the food and rural system in ways which seem to make them more 'hygienic' and environmentally safe spaces of production and consumption.

These trends are most noticeably associated with the regulatory aftermath of the spreading BSE affair with, for instance, national governments and Brussels putting into place highly complex 'safeguards' into the regulation of key places in the food supply chains of beef and lamb. These, however, have the effect of ratcheting up the regulatory costs of the small producers and food processors, and tend to reduce market entry into the most lucrative supply chains (often linked to the main corporate retailers) to those larger producers and processors who can more easily meet the demanded quality criteria which are now set. This further exacerbates the market failure associated with the agro-industrial model by compounding it with higher, hygienic regulatory costs.

Somewhat ironically, however, these costs disproportionately effect those producers and processors who may have been marginal to the causes of the health risks in the first place (e.g. less intensive producers and small abattoirs sourcing locally). These innovations in supply chain management and regulation are deemed to be in the public interest, but from a rural development and, indeed, an environmental view, they

nature is actively constructed through both social and natural actors and institutions. From a rural development perspective, however, it still provides more questions than answers (see Irwin 2001). It tends to downplay the actual power relations in which actors operate, e.g. in food supply chains; their sources of power, and too readily accepts that nature is both constructed and then *can* act. The BSE crisis has been a classic case of how nature can 'hit back' but it has done so in particular social, political and economic circumstance and in specific ways which have been strongly associated with the described hygienic- bureaucratic mode.

A major problem with these perspectives has been to generate the very analytical focus to which many of its advocates espouse. It still needs to begin to provide a real and more substantive micro-analytical account of how social and natural struggles interact in particular places and at particular times. Much of the discussion has, rather, focussed upon the broader, aggregated analysis of such areas as alternative social food movements (such as organics); or the role of agro-industrial science in shaping natural interactions.

In short, these perspectives have yet to link with the realities and materialities of natural agricultural ecologies *in situ*. They have yet to really respond to Norgaard's challenge of exploring the *co-evolution* of the natural and the social in agro-ecological relationships (Norgaard 1994), or test how such relationships may indeed be constructed. A major question remains: *if the materiality of nature is constructed, then how and who (which local actors, which experts, which politicians) should be involved in reconstructing it given the damage now exhibited?* So far, this question is far too normative for social constructivist rural social science; and there is a sense that it is just too risky a business to go beyond the growing identification of diversity, rather than explore how that diversity is created out of social and political struggles. Yes, the current state of agricultural nature in Europe is an active and constructed thing; but how can these octant energies be given a more sustainable direction?

Beyond the obstacles: social nature in rural development processes: exploring the agro-ecological perspective

To build upon, rather than completely deny social constructivist perspectives, I contend, requires the need to creatively integrate the agro-ecological approaches. Agro-ecology emerged in the 1980s as an interesting attempt to establish a scientific basis for alternatives to industrialised agriculture, that avoid the resource degrading tendencies with which the latter has become associated (Hecht 1987). The agro-ecological framework draws upon different intellectual traditions and disciplines, including peasant studies, ecology and environmentalism and development theory. While initially the theory of agro-ecology was mainly developed in the context of agricultural development in the South

(Altieri 1987), in recent years it is spreading to the USA (ref. Gliessman 1990) and Europe.

I would argue that the agro-ecological perspective is highly promising to give a much clearer normative direction to the analysis of social nature relations in rural development, while also recognising the very active and symmetrical nature of nature. Guzman and Woodgate (1999) synthesising the literature on agro-ecological approaches, as well as pointing to the distinct differences between its framework and the more restrictive 'farming systems' approaches, identify nine key parameters as key elements of the agro-ecological approach.

Crisis of modernity

Agro-ecology implies an alternative definition of sustainability from which is generated an ecologically – rather than industrially oriented discourse. A central concept is 'co-evolution'.

Co-evolution

Unlike simply 'symmetry' as proposed by social constructivist thought, agro-ecology refers to the reliant co-development or co-evolution of society and natural factors (Norgaard 1994). It is recognised that farming systems essentially result out of co-production i.e. the ongoing interaction, mutual transformation and dependency between man and nature, that is, between the social and the natural. It is through co-production that both the natural and the social are unfolded into particular forms and relations (see also Van der Ploeg 1997). The agro-industrial model has thoroughly changed the nature of co-production and disrupted many interdependencies between the natural and the social, thereby reducing the renewability of both. The question becomes how can reversible processes be put in place to regenerate interdependency over time.

Local farmers' knowledge systems

Local peasant or indigenous knowledge needs to be seen (both in the North as well as the South) as significantly different from standard scientific knowledge (and its 'hygienic' equivalents – see above) in that they are embedded in local ecology and encoded in culture rather than theoretical and abstract notions. Science is context independent, yet agriculture is actually defined not just by its bio-physical context, but also with reference to its localised socio-political elements. It is based upon the interdependent accumulation of local natural and social resources, practices and knowledges. These are not just about maintaining 'old cultures' but they involve the creation of defence systems to enable a constant replenishment of knowledge systems in the face of the dominating 'eco-technocratic discourses' associated with globalisation.

Endogenous potential

All agro-social systems have their own endogenous potential; but it is a matter of how these get articulated and valorised through social and political processes to what extent these actually are effectuated. The social aspects of endogenous potential refers to local knowledge systems, but also to struggles of local groups to resist, propose and actively construct alternatives to industrial modernisation and to their capacity to develop social networks to enable these. The ecological dimensions are to be found in the promotion of diversity of agro-ecological systems and a strengthening and valorisation of local ecological specificity.

Collective forms of social action

Social reproduction needs to be realigned (and historically in many instances did coincide) with the ecological management of biological systems (Gliessman 1990; Toledo 1989). This involves the construction of collective forms of social action which attempt – either explicitly or implicitly – to challenge the industrial modes of production, consumption and circulation. The social construction of new patterns of circulation involves the (re-)building of networks that, to varying degrees, are composed of citizens of both the centre and the periphery, the urban and the rural, of producers and consumers. This can occur across and through supply chains as it can through different types of space. Agro-ecology thus is founded upon new forms of associationalism.

Systemic strategies

The social factors that are at work (e.g. ethnic, epistemological, ethical, religious, political, economic, gender-based) cannot be seen in isolation. They must be interrelated within an overall understanding of society. Also the broad range of bio-physical factors such as water, soil, solar energy and plant and animal species must be conceived according to the ways they interact not only among themselves but also with social factors. This involves an understanding of energy, material, money and knowledge flows generated in processes of production, consumption and circulation within and between systems. Ethical decisions are at the heart of progressing more systemic sustainable systems.

Ecological and cultural diversity

Agro-ecology aims to not only celebrate cultural and natural diversity (as the post-modernists advocate) but also to progress it and materialise it in new social and natural forms. This means abolishing the notion of a dominant ethnocentric agricultural development model and accepting and promoting diverse pathways of development built upon cultural and physical complexity and richness. This marks agro-ecology out as distinct from both the globalisation logics of the agro-industrial model and its bureaucratic-hygienic counterpart, but also from the diversity 'for its own sake' approaches of the post-modernists and constructivists. It also calls into question the notion of a single, detached price driven market for foods. Social progress, and the progress of agriculture have to be realigned in ways which maintain this diversity.

Sustainable societies

These parameters of agro-ecology begin to re-position agriculture as a key driver for achieving sustainable societies more generally. Rather than a further segregation of agro-food production from rural society, agro-ecology proposes to reinforce interlinkages and foster the multifunctionality of farming systems. As Guzman and Woodgate (1999) argue:

> '*Sustainable societies can only be constructed on the basis of sustainable, locally relevant agricultures ... implying a complete rejection of the homogenising tendencies of the neo-liberal, global modernisation project and the re-direction of co-evolution towards more sustainable ways of living that are based upon the endogenous potential of an infinite diversity of locally relevant agro-ecosystems*' (p. 304).

Agro-ecology and political economy as a tool for theorising rural development practices

No one would suggest that these principles can be straightforwardly superimposed upon any one agriculture or any one agrarian space. Their resonance has, thus far, been mainly applied and explored in a Southern regional and rural context; and has so far insufficiently incorporated the agro-food supply chain realities. However, these principles can begin to be used as important assessments of how far current developments in alternative rural development practices in the European context represent sustained moves towards a more agro-ecological system. Do the current multifarious struggles adopted by farmers across Europe in the face of the crisis of the agro-industrial model add up to any form of agro-ecological break with the past? Do they engender some of the features of the agro-ecological model? Do they represent viable pathways to sustainable agricultural and rural development and, indeed, deserve more state encouragement? Or, alternatively do we see them as merely partial compromises with the agro-industrial model, struggling to create new markets which may disconnect and fragment the social and the natural in new, more pervasive ways?

A central part of the assessment of how far a more agro-ecological rural development paradigm is indeed emerging, should be theorising rural development practices in Europe in reference to detailed case studies and wider empirical research relating to their impacts. Also we need to better understand to what extent these practices may gain momentum in ways that significantly challenge the agro-industrial, hygienic logics identified above. It is only by doing this that scholars will be able to assess how aspects of social nature become more effectively embedded in rural development.

The most recent hygienic mode adopted by much of European food policy has tended to break or at least fracture the environmental question into specific boxes; making it more difficult to make the above holistic connections. Past state intervention, even that pertaining to the environment, has tended to discourage the bringing together of many of the connections necessary for agro-ecology to flourish. This has represented a pervasive form of *constructed marginalisation* over much of the past twenty years in the European context. As a result it has been difficult for key actors to harness the spatial, natural, regional and knowledge based resources necessary to progress *real* rural development. This has led to contradictions and to rural development being seen as a contradictory set of practices which only (at best) score partially on the Guzman/Woodgate criteria/ principles. One possible reason for this is the current priority national governments now give to reconciling the demands and risks of the 'careful- consumer'. In the agro-food sector this is largely done by assembling the bureaucratic-hygienic apparatus- it itself something of a new compromise between governments and capital- in order to stave off further and deeper consumer-led legitimation crisis in the industrial system as a whole. In this context 'primary producers'- those nearest to the natural land-base- are continually disempowered. This has most recently been exposed with the British governments' responses to the foot-and-mouth outbreak (2001), where regulatory controls on the movement and selling of livestock is hastening the demise of the smaller farms through the closure of livestock markets.

While then the conditions for alternatives become even more difficult, this does not stop more grounded and ecologically distinct initiatives from being pursued (see Marsden *et al.* 2001b). What we see then are actors in rural development having to struggle with forming new- spatially-grounded- sets of associations and linkages at a time when European State policies, allegedly in the form of environmental actions, are indeed tending to regulate nature through highly bureaucratic, fragmented and legalistic means. This does not stop the actor struggles from continuing and diffusing across Europe. Indeed, our recent evidence shows the proliferation of new associational initiatives amongst producers and suppliers to create and reconstruct supply chains and networks which reconstitute the role of nature, region, locality and quality in ways which begin to align with some of the agro-ecological principles outlined above. However, a key finding is that new forms of state intervention and regulation are required. A revised political economy is needed. This needs to be reconstructed in ways which empower different sets of actors to re-engage with the social and natural means and conditions of food production and consumption.

This will require that new regulatory styles will need to be established. As with the agro-industrial mode and its hybrid hygienic equivalent, this cannot be left to the private actors alone. It too needs a supporting social and political mode of thought and action.

This will mean that both from a theoretical and empirical point of view real ecological modernisation in the agro-food sector will only be progressed if such an agro-ecological political economy can begin to be reconstructed and adapted.

This is no easy task, but it does give environmental social science a key academic and policy-relevant agenda which brings together sociological, geographical, economic and political science disciplines around questions of social nature. It requires a critical interpretative and a reconstructionist social science agenda which tackles key developmental issues. As McMichael (2000) pervasively puts it:

> '*The crisis of development is a crisis of institutions of governance also. The globalisation project seeks to strengthen states and multilateral institutions as corporate entities but weaken boundaries – both national and, in this case, biological. The counter-movement, on the other hand, is exploring alternative political and social forms – from global citizen networks to community-level organisation to what Wolfgang Sachs (1992) terms:*
>
> '*cosmopolitan-localism, that is local activism situated within its world-historical context.... Much is made of the alternative scale of action of contemporary counter-movements. However, given the scenario outlined here, I believe that such movements cannot avoid engaging with policy-making institutions, and addressing the transformation of states as corporate entities- this process is progressively shrinking democratic political space*' (p. 31).

This chapter has begun to identify some of the pathways and the scholarly obstacles involved in progressing such a project.

9 Understanding agrarian-based rural development

Introduction

In this chapter, I wish to begin to assess to what extent the more recent *rural development dynamic* is taking hold (with reference to England and Wales), and to examine how agrarian-based rural development is being constructed – through the responses and adaptations of producers – out of the overall crisis in the agro-industrial model. It is important at the outset here to recognise that the pervasive dynamics analysed in this book are shaping the governance of nature and European rural space in contested ways. They stem from different origins and they owe their existence to sets of different social, economic, political and academic/scientific constellations. The *agro-industrial model,* for instance, is upheld and reinforced by an alliance of agricultural economists and biological scientists who are still legitimating the intensification and scale economies imported from the American and Australasian modernisation project (see Drummond *et al.* 2000). The *post-productivist project* is alternatively associated with the social geographers and planners concerns of rural capacity and the regulatory mechanisms needed for dealing with the 'consumption countryside' (see Lowe *et al.* 1997). This partly stems from the growing hegemony of what Keil (2000) has called the 'suburban frontier', whereby, rural resources and landscapes are increasingly shaped by a hyper-preservationist suburban culture which needs to redefine itself with nature in a highly particularistic and exclusive way.

The *rural development dynamic,* by contrast, while originally having its origins among the largely 'bottom up' initiatives associated with integrating and empowering 'peripheral' rural communities, is now a much broader and more diffuse church. One which may be able to re-centre the significance of agricultural practices and social ecology (see Light 1998). It holds a potential to re-ignite the potential social and environmental role of agriculture as a major agent in sustaining rural economy and culture. For as Kropotkin foresaw at the dawn of the industrial agricultural phase: 'land is going out of culture at a perilous rate'; implying the 20th century growing severance between *ager* and *cultura*- field from farming community, and nature from human society. Bookchin also recognised this trend; seeing it as requiring a radical alternative which would be an extension of early farming practices, belonging to a different type of moral economy; one which views land as the *oikos* of living bacteria, insects plants and animals. This is a direct contrast with the progress of agri-business (the agro-industrial model), which treats the earth as a resource to be exploited through the denial of its cultural significance (see Macauley 1999).

While these three dynamics are theoretical and conceptual postulates which generalise the contested development trajectories currently underway in rural regions in Europe, it is also important to begin to find ways of relating these dynamics (taking a realist analytical stance) to there material expressions in rural space. Not least to see how they play out o the ground, and how they begin to create their own adaptive geographies of rural regulation. This chapter looks, therefore, at some of the struggles the land-based sector is engaged in constructing a rural development pathway; and posits how this might develop a new regulatory authority in the British case at least.

Recent literature on the rural development dynamic across Europe has converged around a set of key themes. These include several points:

The rural economy does not stand alone, it is part of a national, European and global economy, and as such needs to make references to these wider spheres of activity (both policies, regulations and markets) in its search for new opportunities and the new paradigm of sustainable development which is struggling to be built.

The 'rural development' challenge today is to reintegrate agriculture back into a diversified rural economy, giving agriculture a bigger contribution to the sustainable rural economy and sustainable development more broadly. In order to achieve this we could argue that within different rural regions the emphasis needs to be shifted away from a pre-occupation with *economies of scale*, toward a more multi-faceted approach focused upon *realising economies of scope*. By economies of scope we mean *the opportunities for those involved in land-based activities to create real economies by increasing the scope and value of production and services.*

Within this context it is important to redefine the role of farms as potential multi-functional rural enterprises which serve a variety of markets which contribute to sustainable rural development.

This chapter attempts to analytically progress these issues (particularly points two and three) in relation to new agricultural and rural land relationships in England and Wales, and what these can contribute to overall rural sustainable development and the particular specificities of development this implies.[1] It will do this in three parts. First, it will outline some key conceptual parameters which need to be adopted in the analysis and progress of agrarian rural development. It will then, second, consider four case studies demonstrating the ways in which local producers are adapting and responding both to the growing crisis in the agro-industrial model and the new pressures and opportunities that a rural development model also brings. Third, and in conclusion, the chapter will ask two broader questions: (i) what are the barriers and opportunities for cumulative rural development based upon the evidence presented; (ii) what types of rural development pathways are likely over the next decade? How will the three models play themselves out in these agrarian spaces? What will be the geography of the new rural development model? These concluding questions raise some important spatial questions about what sort of agrarian structures are required or desirable in the pursuit of rural sustainable development; and the degree to which state intervention could or should encourage their development.

Some key conceptual parameters in rural development

What is the farm business: new associationalism for new exchange

To begin to answer some of these questions it is necessary to re-interpret and situate the actual social character of the farm business. Farm businesses are now increasingly located in highly complex sets of institutional, market and regulatory relationships. These have become more complex as the crisis of the agro-industrial model has grown. For instance most farmers have been placed on a technologically driven cost-price squeeze which has meant them racing towards reduced prices in the standard agricultural product sectors. CAP policy and in the supply chains of (the increasingly powerful) corporate retailers have added significant regulatory burdens, not least on market entry on both farmers and small food processors. Moreover, the growth of the post-productivist dynamic has pulled many farmers in other, more volatile local diversified markets, tying them to the vagaries of the fickle countryside consumer. Meanwhile, in many peripheral upland rural areas we have seen the growing interference of the state in attempting to create more diversified types of rural development (such as the Objective One and LEADER initiatives operating in parts of England and Wales). These 'external' pressures and contexts are even more spatially defined such that they condition different types of response not only on the individual farm, but also between rural regions.

There are, currently, a set of key contextually-defining characteristics for these new and complex relationships. These mean that we need to re-conceptualise the role of 'the farm' as a site for production, reproduction and exchange (see Marsden 1991).

Farm reproduction strategies have to navigate a pathway through this complexity in their attempts to maintain reproduction of the business unit, which is often regarded as a key element of rural economy/social fabric (family farm).

Figure 9.1 illustrates what we term the 'new associationalism' which is both generating opportunities for, and driving the process of rural development. Three spheres of activity are identified – *exchange, production and reproduction*. The sphere of reproduction relates to the reproduction of the farm unit, and is conditioned by the combination of land, capital and labour (usually family labour). Through production and then exchange, farm businesses acquire the capital (returns) necessary to perpetuate continued reproduction. When crises occur and the income from exchange (via production) is greatly reduced, farm businesses have several options and strategies which can be employed to maintain reproduction of the farm business. These are varied and include the exploitation of family labour, the sale or leasing of assets (land or quota), diversification into non-agrarian forms of production, cross subsidisation of farming enterprises with off-farm incomes, and so on.

The market does move forward and many new opportunities do exist for farm businesses who are able to innovate, and take advantage of new consumption trends and practices (from high quality food to rural tourism). Rural development policies have a critical role to play in helping farm businesses develop and capitalise on new forms or processes of on-farm production so that they can extract more value from the sphere of exchange (the market), in a way that is able to facilitate sustainable forms of reproduction. The sphere of production has always been diverse, but the opportunity and time to capitalise on this diversity has now increased. The market is beginning to recognise and value a much wider range of farm production outputs, including environmental amenity, food quality, and access. Rather than interpreting the farmer as an 'out-worker' of some greater all-controlling food production system (as in the agro-industrial model), we need to recognise that new opportunities and challenges actually place more control and responsibility on farmers, who now have an albeit variable ability to extract value from rural development processes and activities. The variability of farmers to respond will be critically determined by factors such as geographical location, individual business and management skills, practical skills, access to capital, entrepreneurialism, and the availability of support (both institutional and financial).

From the point of view of the farm then we can see that the spheres of production and exchange are now far more potentially diverse. Production is now more aligned to exploiting the resources of the farm (land, labour, capital, buildings, knowledge bases) in a more 'multi-functional' way, such that the provision of services on the farm may complement the production of goods (including food products of varying degrees of quality). Exchange, and particularly the rules of engagement involved in entering into exchange are associated with a range of entry criteria and qualification associated with the priorities placed by supply chains, regulatory bodies, new policy instruments (such as the proposed modifications to Less Favoured Area payments and the new rural development regulation), and quality of delivery criteria associated with those customers of farm based facilities (such as the 'discerning tourist').

In the sphere of production we see new patterns of labour (both domestic and industrial) having to be developed; in the sphere of exchange we see new skills and risk-taking needed in order to participate in market relationships. In addition, whereas the conventional industrial farmer supplied distant markets through local agents and livestock marts, now he has market relations of differing 'lengths', with some markets coming to the farm, and others involving new networks of actors and agencies beyond his traditional reference group (e.g. farmers markets, retailer-led local supply chains). While these are by no means new processes, in that the development of new forms of pluriactivity have been around for some time (see Arkelton Trust 1993), we now see significantly more structural change and uncertainty in the exchange networks of farms, both within the traditional areas of food supply and regulation (see Marsden *et al.* 2000a) and in the supply of services and new markets associated with the consuming countryside (Marsden 1999). We argue here that this represents a radical break from the past period, when farm diversification was still seen very much as an 'optional extra'. Now, what we are witnessing is more complex in that it involves two multi-dimensional processes.

First, farmers have to develop a set of compromises with these new interfaces and networks whether they like to or not; that is that they have to position themselves, of more appropriately continue to reposition themselves, in relation to these wider spheres. This means that decisions not to engage in some interfaces will be just as important as those where positive decisions to engage are taken. Also it means, given the growing volatility of many of these markets (as much to do with changing regulation as to do with economic competition), that these 'choices' become more critical and not easily reversible.

Second, such new strategies have to be only partially engaged in, such that the farmer and the business can exercise the advantages of 'multi-functionality'. That is the farmer needs to keep more productive and exchange relations operating all at the same time. For instance, the decision to enter a quality food market needs reconciling with producing 'bulk goods' which need reconciling with running a farm shop or bed and breakfast facilities. The synergies needed to conduct these activities *all at the same time,* and /or to balance one off against another is a fundamental feature of this new 'multi-functional' phase'; and it is far removed from the monocultural productiveness of industrial agricultural production.

These multi-dimensional features of production and exchange (and not least their effects upon reproduction of the farm family – not within the ambit of this chapter) demand new skills, knowledge bases for farmers. The case studies below indicate how critical these features are. And their ability to apply these will be critical for the overall success of rural development. In addition, they are, as we shall see, based upon what we might call a 'new associationalism', whereby success in exchange and in production spheres is now much more reliant upon optimising new social networks.

Figure 9.1 Reproduction – production – exchange relationships

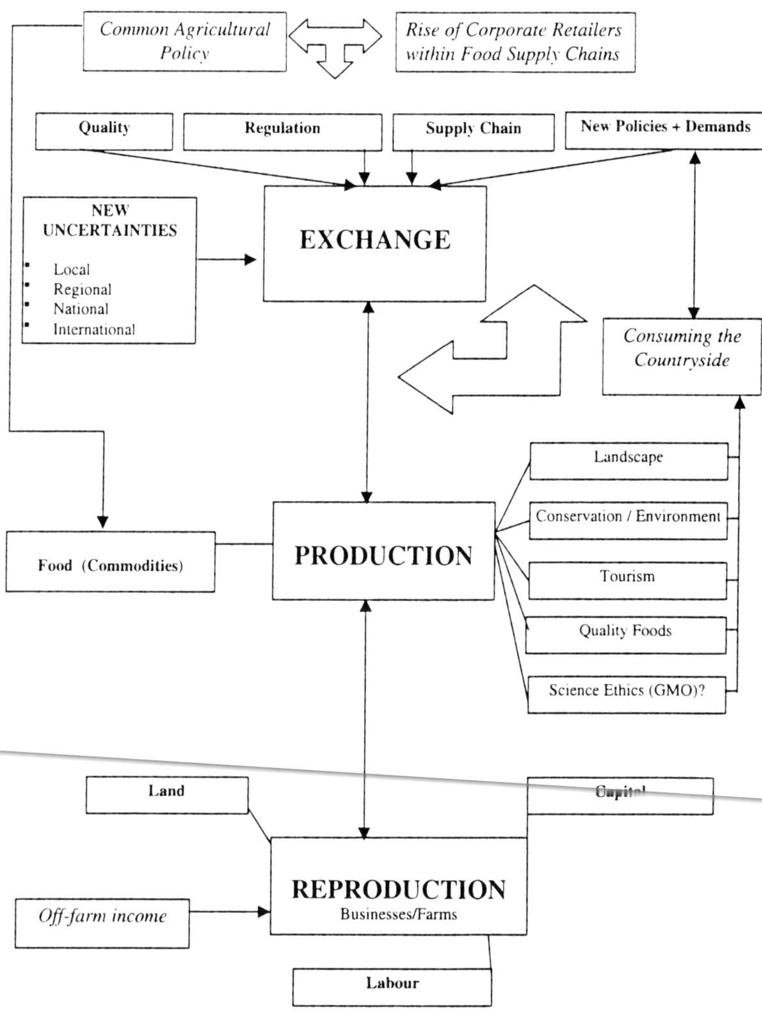

What is (agrarian-based) rural development: from scale to scope

The idea of rural diversification has been around for at least twenty years and has focused, in large part, upon strategies to keep farmers on the land. The success of policy support in this area has been patchy, with many of the most successful developments resulting from the innovative and entrepreneurial activity of individuals operating without any significant public sector support. However, the scale of the current challenge – to reconfigure the structure of the farming industry – now requires a new concerted and co-ordinated approach. It is important to recognise that while new rural development options are developing, many of these are being held back by the continuance of the old-style CAP model of policy support which perpetuates distorted markets based upon single agricultural commodities. New rural development options (as outlined below) require innovation and entrepreneurial strategies suppressed by CAP driven policies, given their emphasis upon providing bulk commodities to highly regulated and universally supported markets. and In keeping with this view, we choose to talk of Rural Development (RD) rather than diversification. So, how do we understand the RD option? And to what extent is RD a non-agricultural phenomenon or does agriculture on the contrary play a crucial role?

Figure 9.2 Conventional agriculture and agrarian-based rural development

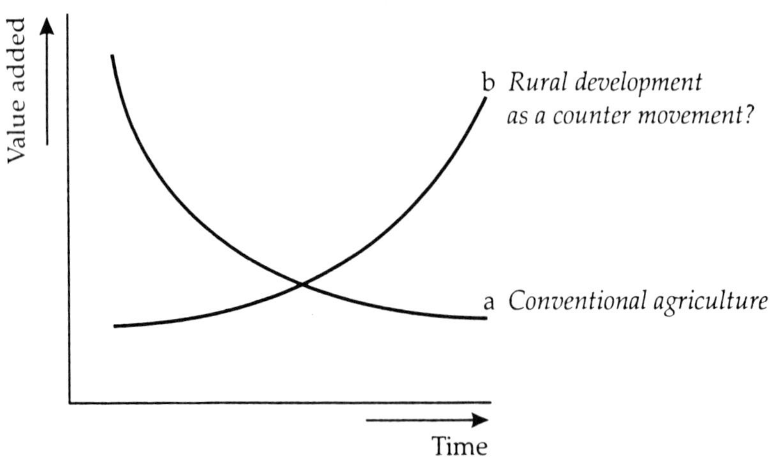

Notes: (a) scale economies in agriculture often mean reducing the value of production; (b) scope economies mean increasing farm-based value added over time

Figure 9.2 illustrates the various conceptions of conventional agriculture and agrarian based RD (i.e. ABRD). The down-going line (a) stands for the 'race to the bottom' in conventional farming as a result of the

price/cost-squeeze mechanism. Line (b) questions whether new RD measures can really deliver a sustainable rural economy through value adding and synergy of operation? Major questions surround the assumptions built into these dynamic trends. What is clear, however, is the tendency for the gradual devaluation of 'farm-gate' agricultural products through both the intense and growing competition on world markets, falling support prices, and rising agricultural input costs. At the same time, the past twenty years has been one of experiencing a gamut of different rural diversification initiatives which have held varying degrees of success. We now witness, as we argue above, the development of a distinct phase; one where the development of real rural development gains through initiating added value activities can potentially reverse the declines in farm income. In aggregate terms this means that agrarian-based rural development (ABRD) can be seen as a search for a new developmental model for agriculture based upon different economic strategies. These include attempting to develop economic strategies in two complementary directions: these are the *economies of scope* and *economies of synergy*.

Both of these new types of economic relations begin to replace the traditional (and increasingly crisis–ridden) 'economies of scale' logics which have dominated post-war agricultural policy and economy and replace it with approaches which begin to evolve the multi-functional, land-based enterprises associated with such activities as agri-tourism, landscape preservation, high quality production. Theoretically, *economies of scope* emerge in so-called 'multi-product firms' (see Saccomandi 1998; Panzar and Willig 1981; Van der Ploeg *et al.* 2000). If farmers can create mutually interlinked products and services which use/exploit the same resource base (be this physical, social, or knowledge-based); i.e. create a synergy between the creation of different value-added products, then this can then lead to scope-economies which are not simply based upon scale. Rather, they are based upon the indivisibility of inputs and outputs which are involved by farm-based resources rather than the progressive divisibility associated with intensive forms of agricultural production.

These economies of scope have to also be linked to two levels of *economies of synergy*: that is those which are associated with the synergies between different activities *on the farm,* and those which are associated with the networks of exchange, co-operation and competition between the farm and the local and regional sphere. Success under these new rural development logics thus means that they need to be created out of different types of social networks or modes of association amongst farmers, between farmers, those involved at different points in the various supply chains, and those significant regional and local actors who are able to facilitate the economies of scope (such as regional development agencies; promotion and certification agents; transaction specialists etc).

Under these conditions we can begin to define ABRD within a context of the following parameters, before then assessing how these operate using some key examples.
To be identified as a 'Rural Development activity' we can postulate that the following conditions must therefore be met:

- The concerned activity is to be, *de facto* or potentially, a response (albeit partially) to the price/cost-squeeze on agriculture. To put it differently: the activity is to be structured in such a way that it *adds* (or will add) income (and employment opportunities) to the agricultural sector. The activity *enlarges* therefore the value added (VA) in the agricultural sector and allows more farms and more on-farm employment to be continued than would be the case without such an activity.
- The concerned activity is to express new relationships between the agricultural sector and society at large. That is, the concerned activity is to contribute to the construction of a new agricultural sector that corresponds to the needs and expectations of society at large. It expresses new forms of 'rural entrepreneurship' (as compared to the previous 'agricultural entrepreneurship').
- The activity implies a redefinition, recombination and/or reconfiguration of rural resources; it contributes to a new developmental trajectory for the rural.

Figure 9.3 Rural development sphere of activity

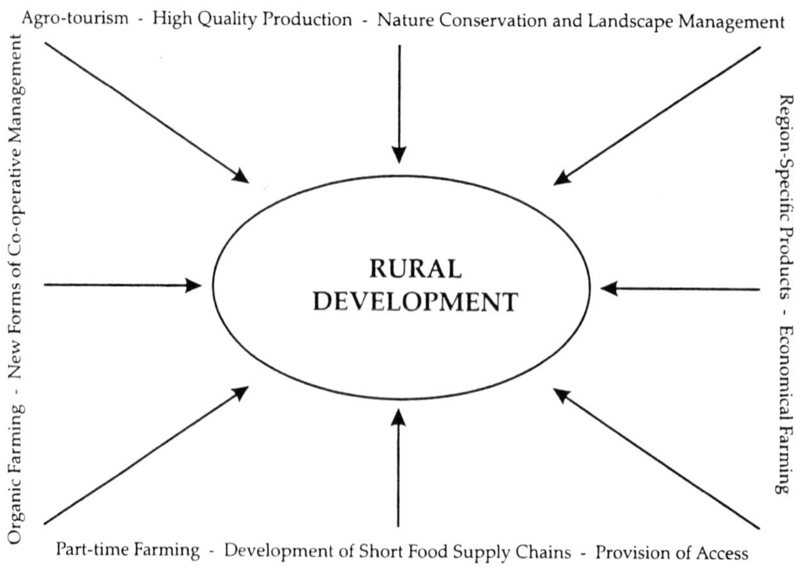

Taking this definition forward it is possible to identify a wide range of RD activities which both in themselves and in aggregate terms leads to more sustainable forms of AGRD over the medium to long term. The following diagram illustrates these.

Realising the potential of the economies of scope and synergy thus depends upon at least the following set of factors:

- Location/ability to utilise location (market access)
- Entrepreneurialism/skills of business manager
- Access to capital
- Freedom to develop new businesses/opportunities within planning /regulatory framework

ARD is thus accompanied by a terrain of uneven development whereby – unlike the previous agricultural paradigm – which sought to compensate farmers for lack of market access/poor production conditions etc., different markets now play a critical role in determining opportunities. There is also a danger, therefore, in considering ARD activities as a rural panacea when they are by their very nature spatial – by definition – therefore often unavailable for some. We will outline in conclusion here how this factor may limit ARD development, and how it needs to be seen in a co-evolutionary context with the other models and patterns of agricultural development.

We outline four case studies here which indicate successful examples of developing *economies of scope* through agrarian-based forms of rural development. It is important here to appreciate that these are exemplars; but that under certain policy-based conditions these may begin to lead the way in developing a more sustainable rural base. We shall, in the final section, focus upon the development of this policy context; making the point that successful *'economies of scope and synergy'* will need to be fostered in their development by a supportive and over-arching policy context which develops a consensus about where it sees rural development going over the medium and long term. The implication is that new rural development activities will not become mainstream unless they are accompanied by a redefined and strategically organised rural development policy framework.

The Cases

Processes of adaptation and response

In this section we outline two case studies which exemplify successful examples of developing *economies of scope* through *adaptation and response processes which result in new agrarian-based forms of rural development*. It is important here to appreciate that while these remain only exemplars, under certain policy-based conditions such activities may begin to lead the way in developing a more sustainable rural base. We shall, in the final section, focus upon the development of this policy context; arguing the point that successful *'economies of scope and synergy'* will need to be fostered in their development by a supportive and over-arching policy context which develops a consensus about the form, structure and function of agrarian activities within the broader rural over the medium to long-term. The implication being that new rural development activities will not become 'mainstream' unless they are accompanied by a redefined and strategically organised rural development policy framework based on a clear, realistic (market embedded) and politically supported, vision of the rural economy.

Civic oriented adaptation and response: the provision of environmental goods and services in Wales

Tir Cymen or 'Tidy Land' was a pilot agri-environmental scheme in Wales which has now been extended nationally under the name Tir Gofal or 'Land in Care'. Tir Cymen was established to test a market-based approach to the conservation of the farmland landscapes of Wales in the belief that farmers have the skills and resources to develop and sell their environmental products and services alongside the more familiar agricultural produce (CCW 1997). The scheme offers farmers an economic return for the positive management of their land for the benefit of wildlife, landscape, archaeology and geology, and for providing new opportunities for quiet enjoyment of the countryside. The voluntary scheme operates on a whole farm basis, with farmers receiving annual payments for following a code of good environmental practice on the farm, with additional annual payments for positive management of important habitats and providing new permissive access. Whole farm agreements also included a schedule for capital works, such as dry stone wall repair or hedge laying and fencing, with capital grants being provide to support the costs of this work. Evidence suggests (see Banks and Marsden 2000) that Tir Cymen has resulted in improved incomes for farms involved (estimated net effect on farm income in 1996 was £1616), and also in a significant increase in the quantity of environmental work on farms (an additional 29 days of environmental work in addition to safeguarding an existing average of 19 days). Furthermore, in addition to direct management payments, 123 per cent of expenditure on capital payments to farmers under the scheme find their way into the local economy (spending on labour and materials), helping to support rural industry and jobs (ADAS 1996). Tir Cymen has thus had a significant impact on the farms participating in the scheme. The whole farm plans have engendered a 'culture of conservation' among farmers, and also enabled many to farm more profitably as a result of improve stock management (resulting from boundary work such as hedge laying and fencing), and reduced stocking rates (improving lambing performance of remaining ewes). As farm incomes have fallen since 1996, Tir Cymen payments have been greatly valued as a source of regular income, and have enabled many farms to remain in business through the current farming crisis. Tir Cymen has also had a major impact in maintaining the environmental assets of Wales. Tir Cymen and its national successor Tir Gofal, illustrate how farmers and policy makers are prepared to adapt and respond to emerging tax-payer/public demands for a well managed, accessible and picturesque countryside. Threats to the environmental heritage of Wales, such as the loss of dry stone walls resulting from the inability of farmers to finance repairs during a period of recession in the industry, have knock-on impacts well beyond the farm gate. On of the most striking features of the Tir Cymen case is that the response to environmental, heritage and access concerns has resulted in a scheme which has the potential to actually increase levels of farm income (resulting from improved stock management) as well as delivering environmental and other public-goods.

Truncating the supply chain as adaptation and response: capturing value-added and reintroducing personal interaction through direct retailing

Nationally, an increasing number of farms are establishing farm shops and attending farmers markets in an attempt to increase the value-added on food production which they are able to capture. For example, the number of farmers' markets has risen from under 10 in 1997 to 4,400 individual markets in 1999-2000, and with an estimated total turnover of £65m in 2000 (NFU 2000). According to Jules Pretty, across the UK farmers receive only £0.09 of every £1 in value-added generated through the food production, processing, distribution and retail of food (NFU Countryside Magazine 2000). Direct retailing is a strategy designed to enable the farmer to capture a greater proportion of this total value-added. For many farms, direct retailing is a major contributor to whole farm income. Survey results indicate that the average turnover of farm shops in the UK is £213,996 although the median turnover is much lower at £90,000 (Banks 2001). The average net income for farm shops in 1999/2000 was £25,049, although the median was again much lower at £12,000 p.a. For those attending farmers' markets (there was strong correlation with those also operating farm shops or other forms of direct retailing), the average turnover in 1999/2000 was £14,466 (median of £3,000) per year. For many small producer-processors farmers' markets and farm shops provide market access, particularly when they may not have the volume of product or capacity to invest necessary to supply major food retailers. While some grants and schemes have emerged to support farm diversification into direct retailing, such activities primarily represent an adaptation and response to new market opportunities, notably the desire for consumers to interact with food producers and for locally produced foods which such short food supply chains are able to deliver (see Marsden *et al.* 2000a). Opportunities are particularly good for those located close to large urban centres. Farm shops bring people into rural areas to shop, as well as providing access to fresh food at a time when many village shops are closing.

Conclusion to case studies

These two case studies show that, with variable degrees of success, it is possible for farmers to increase the level of added value from food production and farming through the realisation of economies of scope. RD activities prove themselves capable not only of generating additional income on farms, but also of supporting a large number of jobs through associated supply chain and supplier linkages. The pursuit of economies of scope is helping many businesses reduce their dependence upon production related subsidy and support, and is working to integrate these businesses more closely and intricately with the surrounding local economy (through consumer and supplier linkages) and environment. The question remains, however, as to how these examples can be built upon, how they can be sustained both over time and space.

New pathways for rural development: scope and synergy in space

The major policy question is, what evidence is there that the types of RD activity illustrated above, are actually capable- in aggregate terms- of delivering the sustainable rural economy and environment over the medium term? How likely is it that such examples will spread such that they become the norm?

To begin to answer these questions forces us to address new, or at least revised structural questions about the types of agriculture which policy – makers and the public wish to progress. This forces us, somewhat uneasily given current debates in the rural sphere, to consider normative options for rural development. For example, what sort of farm structures do we want (in regions like England and Wales), and how can the processes of RD be supported by the public sector to help to deliver this outcome? These are, of course the very macro-economic questions which were raised and answered during the productivist period of agricultural development, and so foresightfully proposed by the Mansholt Plan.

Now, we can argue, is the time for a redefined form of macro thinking based upon the encouragement of diversity and scope of the rural economy, and a re-engaged role for the state in its support. This would really begin to add weight to a sustainable rural development dynamic. The danger with the examples outlined above is that policy makers and scholars alike make the incorrect assumption that they can be sustained and spread without state involvement or support; and that they represent the free-market and post-modern shift towards the stimulation of diversity and difference. This is far from being the case. They are, by and large, highly materialist struggles by different groups of actors who come together in the face of uncertainty and adversity, so as to try something different in order to survive. As we have indicated above they are highly context dependent and unique. The challenge becomes how best to develop policy frameworks and conducive regulatory frameworks which can assist their development. Without this support our evidence suggests that such cases can remain interesting exemplars rather than major drivers for rural development.

Figure 9.4 Potential pathways for the business

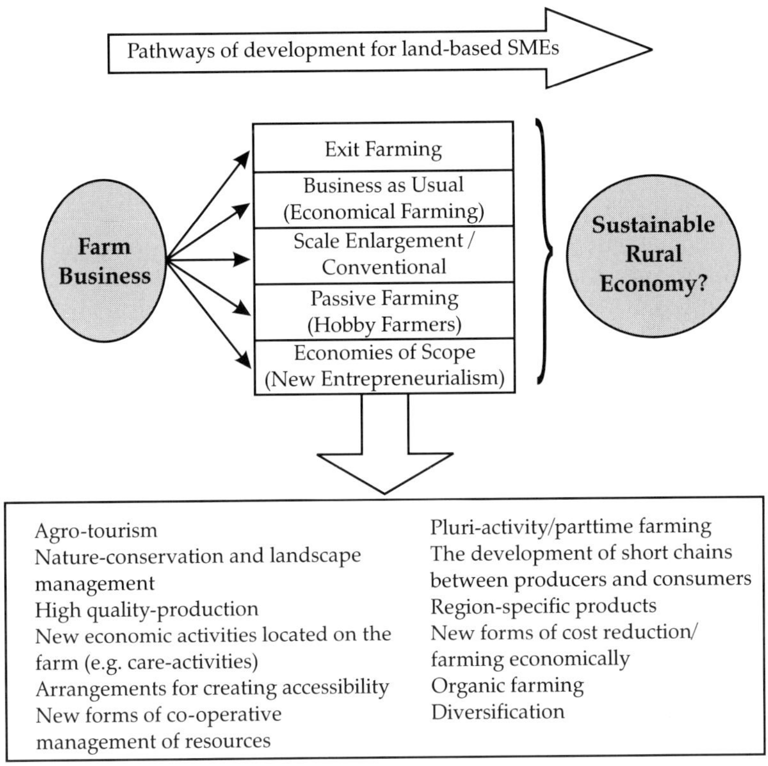

In order to progress policy development in this regard it is also necessary to see 'economies of scope' strategies in the broader context of the range of pathways for agricultural land-holders (see Figure 9.4). The eventual 'holy grail' of a sustainable rural economy will only be achieved by appraising and influencing the structures which can encourage the range of pathways outlined in Figure 9.4.[2]

Currently farm businesses in England and Wales face a number of options or pathways of development which the current crisis of 'productivist support', associated with the agro-industrial model, have placed into sharp relief. Looking forward for the next decade structural changes in the production sector are bound to take place, and the main policy question becomes how can these be influenced so as to progress a more sustainable rural economy. More farmers are likely to retire and give up farming, not having either the energy, time or commitment for economies of scope. With the average age of farmers in Wales currently at 57 this means that the next decade will see major occupancy decisions being made, despite the fact that many of the smaller farmers have not been viable enterprises for over the past two previous decades. The large degree of owner–occupation of small farms has meant that whatever has happened farmers can still sit on their land-rights.

However, inevitable cohort-progression, a growing reluctance for many sons and daughters to succeed the agricultural business, and the increasing demand from ex-urban groups for rural land and buildings, means that more will exit over the next decade. The number of full time dairy farms, for instance, halved between 1977 and 1999 in Wales. These trends are likely to continue with the exit land being re-occupied by either the larger productivist farmers or the ex-urban (cash-rich) lifestyle or what we might term 'passive' farmers. We can see here how both the agro-industrial and the post-productivist development dynamics are juxtaposed in space. Some farmers will attempt to struggle on adopting a 'business as usual' strategy. But this will increasingly be more difficult to sustain given the growing conditionality of former automatic subsidy structures (such as Less Favoured Area Payments) and the sheer economic necessity to supplement agricultural incomes base upon the production of low-value/high bulk products (such as milk and red-meats). The ability for these farmers to reduce their input costs still further (adopting 'economical farming' strategies) will also be curtailed as the input suppliers (machinery manufacturers, seed and pesticide producers) concentrate their attentions upon the larger producers. As a result of these trends, and given only forecasted marginal changes in extant policy structures, it would seem clear that the future occupancy structure of Welsh farming will increasingly shift towards the three rural development models: agro-industrial, post-productivist, and rural development.

First, we will see the maintenance of a profitable conventional agricultural sector based upon scale economies and opportunistic scale enlargement when and where this can occur (depending upon the rate and location of farm exits). This pathway will only develop where quality conventions have been formalised with the wider national and international processing and retailing sectors. It is important to recognise that scale enlargement may not occur to the exclusion of other activities (e.g. associated with the economies of scope). However, it is likely that larger farms will rely most upon income from agriculture for their overall viability. These are the farms which, in order to maintain viability, are able and prepared to increase the scale of operations and thereby spread their fixed costs over a larger number of hectares or animals. They will be characterised as the more professional farmers with dairy herds over 150 cows or 1m litres, sheep flocks over 2000 head.

Second, we can expect a growing colonisation of agricultural holdings by ex-urban middle class groups as a result of the regional and sectoral disparities between the housing and agricultural land markets. For this reason, and given the reductions in the ages of retirement, more ex-urban groups are going to wish to buy a slice of rural land as a more extreme expression of the suburban lifestyle. This is already occurring at a fast rate in parts of middle England and it will spread wider into the more peripheral rural regions of Northern England and Wales. This means that agricultural land (and especially) buildings prices will be kept high despite the continued falls in agricultural incomes.

These are farms where 75 per cent of whole farm income is derived from non-agricultural sources. Current statistical definitions of part-time farming tend to under-estimate the number of these farms ; and the farm business may well operate at a loss but be cross-subsidised by other income sources. Hobby or leisure farming is impossible to quantify from current statistics, but anecdotal evidence suggests that they are on the increase and that they rely upon at least one member of the household generating significant amounts of off-farm income. We can expect this pathway to increase over the next decade as the demand for small holdings (including farm buildings as well as some land) grows.

Third, we can also expect, as our case studies suggest, the growth of the 'economies of scope' rural development strategy where farmers become far more entrepreneurial, and begin to transform their agricultural occupancy rights into diversified or multi-functional land-based SMEs. This strategy, as we can see above can have a range of different expressions: including expanding other on-farm income (e.g. tourism, retail); capturing greater value-added in production via an involvement in processing; and/or diversifying away from commodity production (e.g. into organics/niche/regional/quality production; or singularly or in combination converting agricultural land into environmental amenity.

These different pathways are already clearly occurring. Their development and sustainability, however, will partly depend upon the types of encouragement and assistance they obtain from state policy. It is clear that most of the positive changes with regard to farm occupancy involve what we might term *a farm-plus strategy*. That is, they depend-as our diagram in the first section of the chapter suggests- upon the development of extra-associational involvement of the farm household with a growing range of outside agencies and actors. A major problem for state policy is how best to facilitate and encourage these given that they rely upon highly context dependent circumstances. What is increasingly clear is that whatever the pathway chosen by farm holders over the next decade, the no change, 'business as usual' option can only be at best a short-term expedient. Similarly for state authorities, strategic action in actively influencing the more sustainable rural development pathways will be required such that the struggles surrounding the development of the micro-economies of scope can be matched with the broader state support for the practices of sustained and new associational market and productive development in the rural, land-based sector. The pathways outlined here, indeed challenge the notion of the 'family-farm' based solely on bulk agricultural production as the centre of the agrarian model of development. It will be progressively replaced with the shift towards leisure and life-style farming, professionally-run entrepreneurial holdings and multi-functional businesses which exploit the economies of scope and synergy. While in England and Wales there is growing policy discussion about how to assist existing farmers and farm structures to adapt to the pressures of world markets ('race to the bottom' scenarios) this has not led to a serious equivalent attempt to grasp the new farm structural questions associated with the new rural development trends now also becoming established. This lack of policy consideration could seriously affect the success of these new trends, hampering sustainable rural development and maintaining the vestiges of an increasingly inappropriate agro-industrial model.

The test of success for the new rural development agenda will partly stem from the ability of the state to match the struggles for the economies of scope with *a strategic approach to the developing occupancy of rural land*. This can no longer be avoided given the growing consensus about rural sustainable development on the one hand and the growing realisation that we need to replace the old post-war productivist strategy with an equally robust and normative model of sustainable rural development. It thus requires both farmers and policy makers to realise that the status quo is not only no longer an option, but it is also no longer economically or politically viable. The analysis here at least begins to show that there is light at the end of the tunnel but that to keep it alive will require more than the detached state support for bottom up rural development initiatives.

By exploring some of the new conceptual parameters associated with agrarian based rural development, and in outlining some of the material cases where this is effectively working (at least for the time being), this chapter has begun to explore the ways in which the future of rural areas *could be socially managed* in more sustainable ways. Increasingly, it would seem that rural space and its actors are having to align themselves to alternative pathways associated with the development models outlined at the beginning of the chapter. The ascendancy of the rural development model is by no means inevitable. It will require shared building of social and political legitimacy in order for it to gain a sustained authority, rather than a form of arrested marginality. Social scientists, in beginning to demonstrate what can and cannot work, have as critical a role in developing this project, just as they have had, and still have, in the fuelling the agro-industrial and post-productivist logics. *Are we up to helping to create a revised moral economy for rural areas, and to help (to infer from Kropotkin) to put back the culture into agriculture?*

Notes

1 The analysis and the empirical evidence upon which this chapter is based stems in the main part from the EU IMPACT study of which the authors are members. The first part of this study has involved the development of in-depth case studies of farm-based rural development initiatives. These are summarised in the case study section of this chapter. The broader analysis is also supported by the authors variable involvement in the reform of rural policy currently being undertaken in England and Wales.

2 Although not completely specified here, the construction of Figure 9.5 and the subsequent discussion are based upon the authors attempts to estimate the key directions for current England and Wales farm occupancy. These estimates are not helped by the continued obsolescence of official agricultural statistics in England and Wales. Perhaps surprisingly it is still as difficult as it was a decade ago to obtain any comprehensive data on multiple income sources of farm households in the UK.

product becomes, the more scarce it is in the market (e.g. the region of Champagne can only produce so much wine). It is important to note that any one business may be involved in supplying one or more of these different food supply chains. The three main types of SFSC we identify are:

- Face-to-face: consumer purchases a product direct from the producer/processor on a face-to-face basis. Authenticity and trust are mediated through personal interaction. The internet also now presents opportunities for a variant of face-to-face contact through on-line trading and web pages.
- Spatial proximity: products are produced and retailed in the specific region (or place) of production, and consumers are made aware of the 'local' nature of the product at the point of retail.
- Spatially extended: where value and meaning laden information about the place of production and those producing the food is translated to consumers who are outside of the region of production itself and who may have no personal experience of that region.

Complexity and competition

The literature on the development of 'alternative foods' is already large (if still highly fragmented and untheorised). Our focus here, which bears heavily upon agriculture-based rural development, is upon the nature of these 'supply chain relations' rather than upon the particular characteristics of the foods themselves. Types of speciality, quality, region specific, or organic foods are by no means solely the preserve of the alternative mode. Indeed, near identical products may emerge from both of these modes of food supply. This is an important distinction to make given the intense and highly contingent competition between the industrial and alternative modes of food supply, as well as between the different retailers and processors involved in the industrial chains. This is producing some interesting mutations with regard to supply chains. For instance, corporate retailers are now developing home deliveries through internet ordering and exploring closer links with their local suppliers (see Guy *et al.* 2000). Alternatively, some of the successful speciality and alternative quality chains have extended into national and international supply links and markets (e.g. Parmigiano Reggiano). This means that producer-customer interfaces are becoming, and are likely to increasingly become, far more complex and diverse; not so much in terms of the products they supply as in terms of the types of relations and organisational features they display. *In addition, and as we shall see in the following case study, these latter associational features become key influences upon the attribution and allocation of economic value across the different actors in the supply chains.* These processes are identified as a significant research gap in recent literature (see Murdoch *et al.* 2000; Murdoch and Miele 1999), whereby the dominance of work on the standardised/generic model of industrial food supply is matched with a paucity of understanding about the operation of specialised and dedicated supply chains (Storper 1997).[1]

From a rural development perspective we argue that to conceptualise and parameterise these emerging social and organisational characteristics becomes a major challenge; for it is through this understanding that we can better judge the extent to which rural actors – whether they are farmers, processors or retailers – can *create additional value for rural regions.* From a survey of short supply chains in Europe we now attempt to draw out some of the key features which help to build up an improved conceptual picture of short supply chains and their particular role in agrarian based rural development.

The dimensions and evolution of short food supply chains

Figure 10.1 highlights just 5 of 12 case studies developed within the IMPACT research programme[2] which focus upon the reconfiguration of food supply chains and the role of these supply chains in the process of rural development. To refine our understanding of the rural development potential of short supply chains we concentrate on their evolution and dynamic characteristics. Cases in Figure 10.1 range from those which derive from an ecological or regional origin to those which have been established in order to 'shorten' previous chains, or to develop alternative chains and networks which provide rural development opportunities in terms of value added. Figure 10.1 arranges these according to the type of short food supply chain they involve and the degree to which they focus upon an ecological or regional identity. (See Table 10.1 for summary overviews of case studies.)

Analysing all of the case studies undertaken,[3] it has been possible to identify some common dimensions of short food supply chains. This is despite the fact that they are, by their very nature (as we shall see below) highly context dependent. Nevertheless, we can begin to see some common features which help us distinguish these new and alternative patterns of rural development.

First, all of the cases are, to varying degrees, removed from the conventional chains associated with the provision of bulk food commodities to complex food chains. Second, the cases tend to display new relationships of association and institutionalisation which are located at the spatial scale of the region (rather than the nation) and or locality. Third, the companies and actors involved in the short supply chains hold different relationships with the state in that they are either developing new innovations which go beyond state support, or they are resisting the negative effects of state policy. Fourth, they represent new experiments and innovations which combine or reconfigure the natural, quality, regional and value constructions associated with food production and supply. The foods produced and supplied in this sense hold composite forms of value which go far beyond the simple commodity value form. They add to this other types of value associated with natural, regional, production quality identity.

Table 10.1 Summary of IMPACT supply chain case studies

Frankfurt Farmers Markets (DE)	The case study 'Farmers markets in the city of Frankfurt' focuses on the establishment of short chains between producers in Lkr. Vogelsberg and consumers in Frankfurt. Farmers are successful if they offer high quality food products and/or products with a particular regional or natural image. Also the farm households that are engaged in processing and direct marketing activities are often particularly innovative; new products are being developed and old products rediscovered.
Ecological Co-operatives in Andalucia (ES)	Farmers and consumers in Andalucía organise themselves, spurred on by a complementary generic aim: the wishes of the farmer to produce and sell ecological produce in Andalucía and the wishes of the latter to consume them at a reasonable price. Amongst the aims are the following: minimise the middlemen to assure good prices for the producers and the consumers; consumers supervise the planning and control of production to avoid fraud; promote regional production and consumption of these products; facilitate the development of local economies.
Graig Farm Organic Meat (UK)	Graig Farm is a meat processing, wholesaling, retailing and mail order business, located in mid-Wales on the border with England. The livestock used by Graig Farm are supplied by members (20) of the Graig Farm Producers Group, most of whom are based in the region. Nearly all of the products sold by Graig Farm are certified as organic.
Rhöngold Dairy (DE)	The Rhöngold development is linked with environmental improvements and positive socio-economic changes. The most important are the improved income prospects for dairy farmers in the region and the increased potential for the development of rural and green tourism.
Parmigiano Reggiano (IT)	The Parmigiano Reggiano production system maintains the artisanal, hence labour intensive, production techniques in cheese-making, which are crucial for the final quality of the cheese. The basic formula of success is to be found in the generation of a collective agreement of the involved actors to comply with the production regulations, which differentiate the production and processing of Parmigiano Reggiano milk from milk produced and processed in industrial dairy systems.
Llyn Beef (UK)	The Llyn Beef Producer Co-operative is a registered company established in May 1997 by a group of lowland beef farmers in north-west Wales. The co-operative was established on the belief that a premium price could be secured by marketing beef produced from the area on the basis of its quality. The co-operative was also designed to enable local farmers to achieve a closer relationship with retailers and other market outlets.

Fifth, the cross section of short food supply chains examined show positive if variable value-added gains in terms of farm-level income impacts over and above those which would have been possible through the commodity-oriented mainstream 'industrial' channel. Finally, there is considerable variation in the types of associational and face-to face interactions (what Storper calls the 'interpersonal world') which are involved in the production, 'animation' and sales of foods in these alternative short chains.

Figure 10.1 Cast studies associated with short supply chain reconfiguration

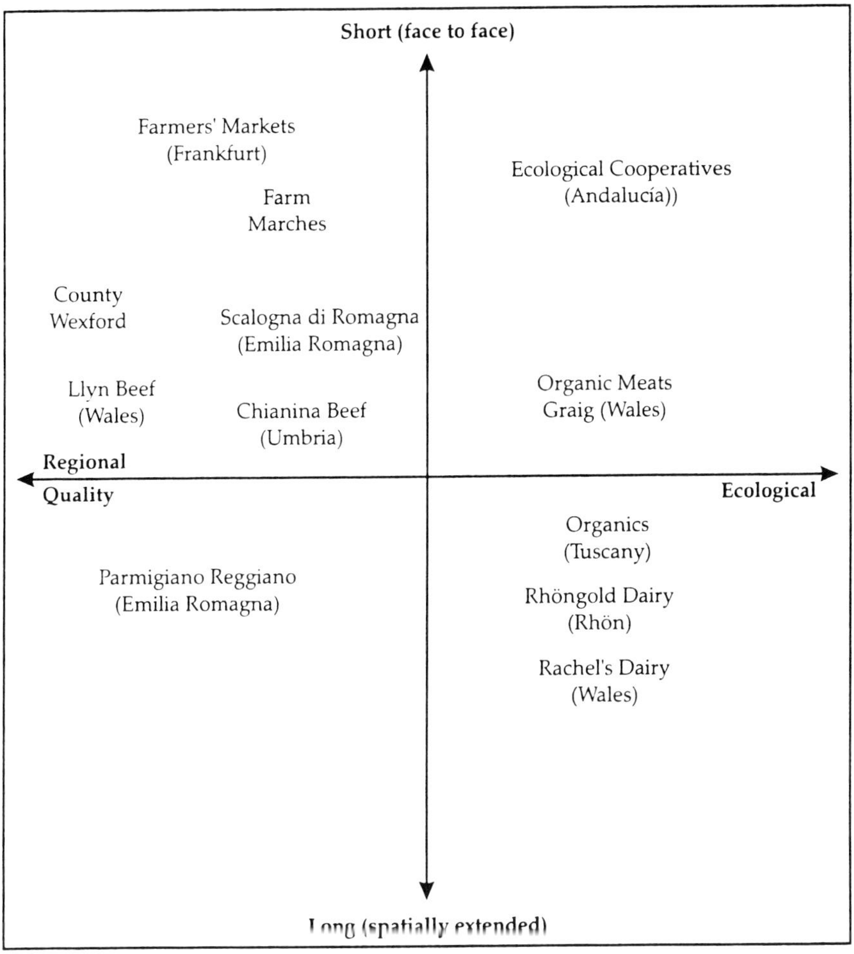

Evolutionary considerations

If such short food supply chains are able to play a significant role in the process of agrarian based rural development, it is important for us to identify and analyse any evolutionary patterns in their development, and then to question and consider their long-term impact and future potential to contribute to rural development as a genuine 'counter-movement' (Van der Ploeg *et al.* 2000). These questions engage us more clearly in beginning to understand 'impacts' and the aggregation of 'little' impacts into clustered impacts (either in the region and/or along the supply chain), as well as the limitations and obstacles to development facing specific types of short food supply chain development.

Figure 10.2 The relative sustainability of short supply chains

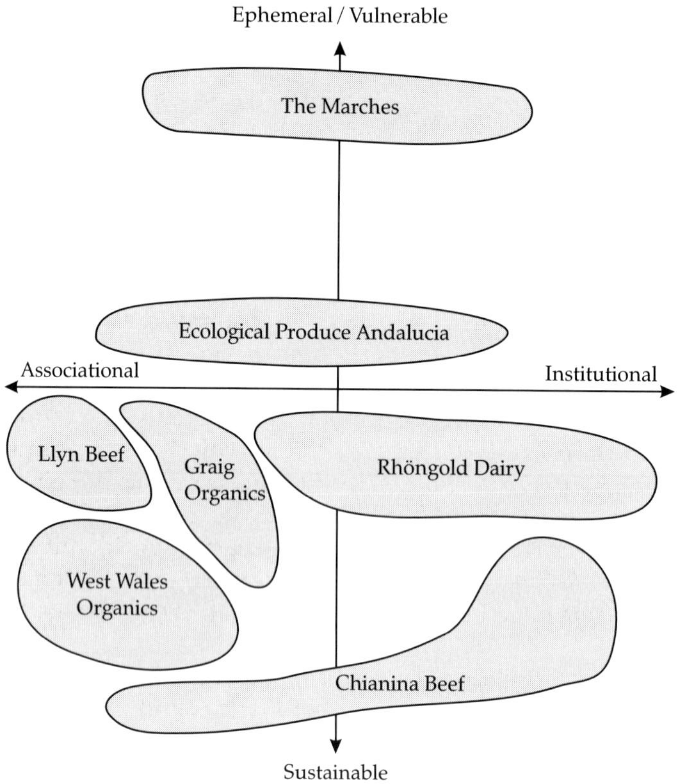

Some food product supply chains are highly ephemeral, others endure generating long, lasting benefits to rural areas. Some supply chains remain highly localised, servicing a dedicated but small group of consumers, others expand to meet consumer demand at a national or even global scale (e.g. Parmigiano Reggiano). Some food supply chains are highly dependent upon associational or institutional arrangements at the local, national or international level, others are less closely interwoven with socio-political structures and are the result of individualistic entrepreneurialism. We believe that there is value in examining how, why and under what conditions specific food supply chains develop, as this is necessary to inform our analysis of rural development potentialities across different regions and production sectors. In so doing, we identify four key parameters of supply chain evolution. Through this categorisation we argue that unlike Storper's static typology (see Figure 10.2) we need conceptualisations which reflect the dynamic and evolutionary nature of supply chains and the businesses they involve (see particularly demand evolution below).

a. Temporal evolution: Reviewing the background to supply chain case studies across Europe, it was observable that two distinct periods of development had taken place. The first has origins starting in the 1950s and 1960s with case examples often receiving institutional backing and demonstrating a strong sense of regional identity. The second period commenced in the early 1990s, perhaps as a reaction to the crisis of the conventional agricultural system and/or the existence of funding and effective support structures (e.g. Rhöngold Dairy).
b. Spatial evolution: Over time, the growth of demand for specific products can provide conditions for the expansion (or continued viability) of certain forms of production within regions. The growth in demand for a Product of Designated Origin (PDO) may transform the economic viability of farmers in that region, especially if a high proportion of a regions productive capacity (land) is required to satisfy demand (the case of Champagne in France as both a region and a product is an obvious example here). Some short food supply chains will have a greater potential to involve many farms and thereby become central to a region's agriculture (e.g. milk production for Parmigiano Reggiano in Emilia Romagne, or champagne grapes in Champagne), although identifying which products are destined to capture larger markets is extremely difficult to do and may depend in large part upon the vagaries of the market.
c. Demand evolution: Often closely linked to spatial evolution (above), demand evolution encapsulates the capacity of markets and distribution channels for specific products to expand from one scale of operation to another, for example from local to regional and perhaps to international levels. Parmigiano Reggiano cheese is a good exemplar of this process whereby a product which once had a very small level of sales outside of the region of production, through marketing and internationalisation of cuisines, now services a global market. Hence as short food supply chains evolve they may extend to span both generic and dedicated food markets, and may adopt 'standardised' features in product development. Figure 10.2 demonstrates how different supply chain cases straddle multiple points in Storper's 'four worlds of production', and that they do not necessarily fit neatly into simple typological frameworks. We are, after all, not talking here simply about producing nuts and bolts. We are dealing with the variable manipulation of a natural product to diversified and ever more health, quality and ecologically minded consumers. In order to maximise the potential of supply chain development, companies and

agents involved may need to adopt a flexible and dynamic approach to product development and marketing, perhaps suggesting the need for a greater degree of foresight and innovation than is the case in industrialised commodity sectors. However, it may also be the case that the ability to expand (or meet) growing consumer demand does not necessarily rely upon rapid or constant product innovation or development, as is arguably the case for Parmigiano Reggiano.

d. Associational and institutional evolution: Associational interfaces (networks) are often informal but are highly significant in establishing trust, common understandings, working patterns, and forms of co-operation between different actors in a supply chain. These differ from institutional interfaces which include state regulations and the support and services offered by economic development agencies. Associational interfaces are often critical in generating and facilitating supply chain initiatives at a regional level (see Llyn case study), however, such interfaces are vulnerable to internal and externally generated disruptions. *In short, there is no inevitability that strong and mutually acceptable associational interfaces will be reproduced over time.* Furthermore, where such interfaces do not exist (or have broken down), it may take many years to rebuild relationships and trusts to a point where regional actors, or actors across a supply chain, can create the conditions necessary to effectively and efficiently meet and maximise consumer demand. Evidence from the IMPACT programme suggests that associational interfaces are both a cause and an effect of the development of short food supply chains. Where interfaces are strong on both it usually provides a strong basis for sustainable reproduction and development over time. The case studies indicate that regional support agencies coupled with deep associational involvement and a professional form of entrepreneurship seem to be critical.

Institutional evolution may radically alter the prospects for supply chain development. For example, the decision by the state to increase support to organic food production may stimulate production and, in turn, enable both product and market development. Interestingly, many of the recent developments have occurred without significant institutional support. *The evidence would suggest that sustaining rural development through the evolution of reconfigured supply chains must be based upon both institutional support and associational development; furthermore, these relationships must alter and reconfigure over time.* Here, concerning these interactions between the farm, the institution and the associational realm, there is no one model. A key question then is the degree to which these features will need to become more widespread if aggregated rural development impacts are to be achieved, and how, if they do not exist, they might be generated? To explore these evolutionary considerations further we now present one of our case studies in more depth.

Case study: The Llyn Beef Producers Co-operative

Concept and genesis: struggles of value

The Llyn Beef Producer Co-operative is a registered company which was established in May 1997 by a group of 40 lowland beef farmers to market beef from the Llyn Peninsula (spelt Lleyn in English) in north-west Wales (see Figure 10.3). It was established in response to the livestock crisis in Wales, which was characterised by falling prices in traditional beef commodity markets and a lack of consumer confidence in the quality of the final product.

Figure 10.3 The evolution of short supply chains according to Storper's typology

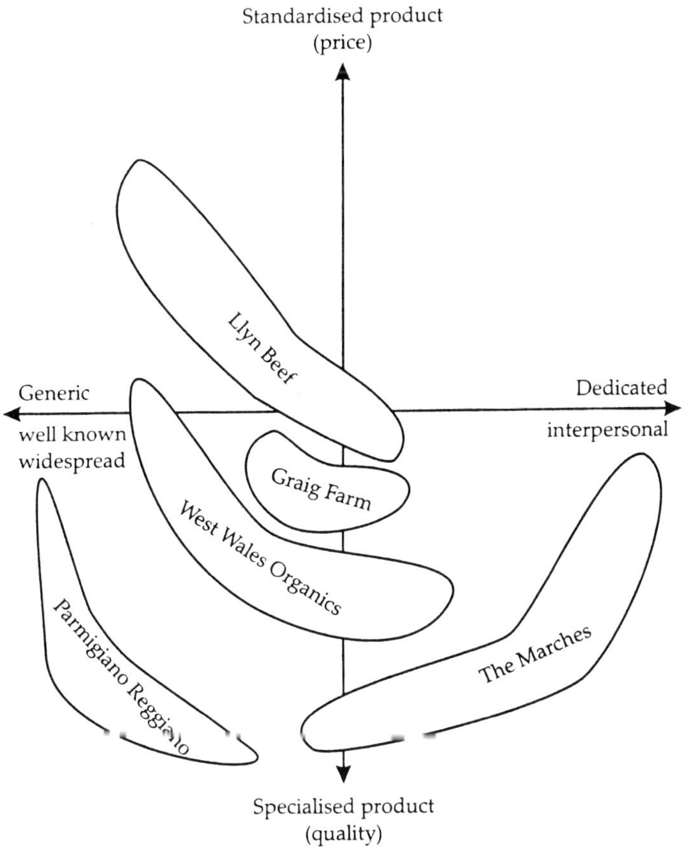

The belief amongst beef farmers on the Llyn Peninsula that '*something needed to be done*' to secure their livelihoods[4] prompted the development of the Llyn Beef Producer Co-operative. The Co-operative was established to try to improve the collective strength of farmers in the supply chain and thereby improve financial returns to beef farmers in the area. It was felt that a premium price could be secured by marketing the beef from the area on the basis of (i) the natural grass-based system of production, (ii) the eating quality of the beef, and (iii) the quality and complete traceability (i.e. assurance or origin) of the product. Added value is generated by co-operation along the supply chain, supported by technical advice (particularly relating to a new beef maturing process), and product marketing and promotion. The co-operative was also established in the belief that a closer relationship with retailers, combined with group marketing, would improve the continuity of supply and reduce the risk of retailers making unrealistic demands. Thus, since its establishment, the co-operative has focused its efforts on securing a contract with a major wholesaler, as well as establishing premium local market outlets.

The Llyn Beef Producer Co-operative represents a response to new consumer demands for quality-assured, fully traceable, mature, lean beef of consistent eating quality. It is also a response to broader societal demands for natural methods of animal rearing, stress-free handling and transportation, and the maintenance of bio-diversity through mixed grazing, extensive management of semi-natural and unimproved grassland, and judicious use of fertiliser and animal manures. The co-operative was established on the premise that the promotion of branded Welsh beef, produced to assured standards of quality is more likely to achieve retailer and consumer support.

Evolution over time and space

Sales of Llyn Beef to local markets began with 2-3 cattle per week to an outlet through the local abattoir, Cwmni Cig Arfon, in May 1997. This quickly expanded to 6-10 cattle a week by September of that year. In the summer of 1998, the Llyn Beef Co-operative attracted the attention of Livestock Marketing Limited (an organisation established to create marketing opportunities for farmers by setting up farmers groups to supply retail outlets). Livestock Marketing Limited subsequently acted as a facilitator for the development and expansion of the Llyn Beef Co-operative. They did this by putting the farmers in contact with both Booker Plc (the largest Cash and Carry operator in the UK), who were looking for a supply of quality, mature Welsh beef, and with the Agricultural Development and Advisory Service (ADAS), who were operating a project in support of quality Welsh beef production.

By September 1999 the Llyn Beef Producer Co-operative had around 100 members, accounting for 98 per cent of beef farmers in the area, with seven farmers acting as company directors. Farmers are required to pay a membership fee to join the scheme and must also sign up to a range of provenance, quality assurance and contractual specifications. The beef produced under these specifications is marketed as 'Extra Mature Welsh Beef' for the wholesale contract with Booker, and as 'Llyn Beef' for local butchers and the domestic (Welsh) catering trade. The main difference in the specifications required to supply these two separate supply chains relate to the provenance.

The supply chain works as follows. Farmers pre-book their cattle with Farmers Marts Limited (the local auctioneer/livestock market), which acts as a procurement agent when supplying to Cwmni Cig Arfon (the abattoir). Although stock goes directly from the farm to the abattoir, grades and weights come back via Farmers Marts Ltd who undertake to pay farmers within one week. The majority of cattle currently traded through the Llyn co-operative fall into the 'Extra Mature Welsh Beef' category and are sold through Booker Plc. Only a small proportion of Llyn beef cattle – about 6 to 8 per week – fall into the 'Llyn Beef' category for sale to local butchers, caterers and restaurants. As with the Extra Mature Welsh Beef, cattle for the production of Llyn Beef are procured through the livestock market but handled directly by the abattoir. The meat is sent out from the abattoir either in primal form (vacuum packed) or on the bone, depending on the butcher's particular preference. Outlets are mainly in the Llyn area, although hotels further afield have begun to express an interest in stocking the product for their menus. The major selling point locally is that beef supplied in this way comes from cattle that have not left the area. This is supported by a 'belt and braces' approach to traceability, whereby traceability cards accompany Llyn Beef at points of sale. These cards detail the farmers name, the name of the butcher or other outlet to whom the beef has been supplied, the length of maturation of the beef, the tag number of the animal and the carcass number. The local Leader II group have been instrumental in both finding local retail outlets for Llyn Beef and providing financial support for the development of customer profiles and publicity/point-of-sale materials. All such materials make strong identification within the Llyn area, but also deliberately make use of Welsh farm-assured labels for generic marketing purposes.

The whole venture has, therefore, been based upon the development of new networks of co-operation along the beef supply chain. As a representative of Livestock Marketing Limited explained,

> *'A number of initiatives taking place around the same time were brought together to form a model of co-operation, from primary producer to consumer, backed up by technical advice and promotion.'*

The benefits of co-operation are felt along the supply chain (see below). According to one of the Llyn Beef farmers:

> *'we have faced one or two difficulties along the way, but these have been overcome through the commitment of all the partners involved, and this is what has enabled us to expand. Farmers are benefiting from co-operation with one another and getting together. We have realised that we need to co-operate to survive and also to handle bulk.'*

The Fresh Meat Controller at Booker Plc has also emphasised the benefits of co-operation:

> *'Our customers are delighted with this product. The key has been the success we have had in developing close relationships, mutual trust and co-operation with our Welsh partners. I am convinced that there is a direct relationship between the level of co-operation and the quality of the end product.'*

Assessing the socio-economic impact

The difference between gross margins on Llyn beef farms and traditional lowland beef producers is chiefly caused by the premium price received on the final product. In other words, while costs are relatively similar, the value of output produced is significantly higher.

The extra value added from Llyn beef farms can be calculated as the difference between the gross margins (£ per head of cattle), in this case £27. The extra value added per farm equals the difference in gross margins per head of cattle multiplied by average size of Llyn beef herds, which equates to £5670 per farm (£27 multiplied by 210 cattle). Multiplying this figure by the total number of farms on the Llyn involved in the Co-operative gives an indication of the total extra value added to farming in the region (£5670 times 100 farms) of £567,000. These estimates are supported by the farmers themselves. One farmer stated that *'for a farm of my size I can secure a premium worth an extra £5000. That is a significant mark-up on the price that could be secured elsewhere'*. It is also worth observing that a premium of £5000 is very significant in the context of current beef farm incomes in Wales.

As well as the extra value-added created on Llyn beef farms, the abattoir (Cig Arfon) benefits from the extra trade generated by the contract with Booker Plc. This is estimated at £2.22 million. The abattoir also estimates that sales under this contract could increase by a further 20 per cent in the medium-term, indicating that there is potential for the impact of the Llyn Beef venture to increase in the future.

Overall impact: towards an evolutionary approach

Clearly the Llyn Beef case is a recent example of the development of short food supply chains prompted by the severe crisis and cost-price squeezes of the British livestock industry. It is a particularly successful case, based upon a somewhat unusual degree of co-operation between producers who do not have a tradition in this regard. The case demonstrates how significant amounts of value added have been generated at the farm level by the initiative, and that new supply chains can be successfully created, even by groups of formerly 'powerless' producers operating in the mass industrial food system. In addition, it is clear that the venture can only be sustained by encompassing other networks of actors who can assist in innovation and support and recognise new quality conventions. The case demonstrates that *synergies on the farm* (through the development of economies of scope and reductions in opportunity costs) need to be complemented by *synergies both between farms and with the other actors in the food supply chains*. In reality two new short supply chains have been created, based as they are upon slightly different rearing and processing conditions, and serving different types of customers. This represents a sophisticated type of new product specialisation operating in these new supply chains. This level of organisational innovation is surprising given the traditionally recognised low base level of innovation and room for manoeuvre associated with the mass industrial chains upon which the Llyn farmers were previously reliant.

The case also shows how the corporate retailing and catering sectors are also increasingly aware of the need to develop value added and 'knowledge-added' lines for their customers, supporting consumers' construction and perception of quality. In Storper's terminology, this represents the development of more *dedicated* marketing based upon more specialised (as opposed to standardised) products. This can only be done, however, through the emergence and evolution of new organisational structures (such as the Cwmni Cig Arfon abattoir and the Farmers Marts developments) *as well as* the development of new co-operative activity between the producers themselves. The development of these short supply chains is, therefore, anything but simple. It requires the building of new socio-economic relationships which operate in different spatial frameworks (in this case associated principally with the Llyn peninsular). Here we can see how new equations between space, nature, quality value and product come together. This does not happen automatically. It demands carefully organised action and association which creates and then *maintains* the new synergies. For them to come together elsewhere, it is likely that they would need to be reconfigured, even if there are some relevant common principles. Clearly, the evolutionary, space-time and synergy processes involved in this case are highly particular. This is so because of the fact that in such new rural development initiatives, uniqueness and distinctiveness at the place of production needs to be matched and articulated forward through to the point of consumption. The resulting supply chain relations are like unique 'forks of lightening'; they occur intensely across spaces in one vector, but by definition they are difficult to replicate in identical form elsewhere. This does not stop other groups of producers, processors or development agencies attempting to replicate the success of Llyn Beef; but it does mean that they will have to forge distinctive relationships and different natural/local equations with different sets of food supply chain agents (including retailers, abattoirs, etc.). This suggests that as far as short food supply chains are concerned, their role in creating new forms of rural development will be to increase the distinctiveness of space and diversified agricultural relations.

Conclusion: overcoming the inherent constraints

Whilst we can see that the farm-based and local impacts of short food supply chain developments may be positive in terms of value added, what is less clear is the degree to which such rural development initiatives can be *sustained and developed both over time and space*. For this to occur we need to progress theoretically the concept of *rural development clustering*; that is, the degree to which initiatives such as Llyn Beef can be built upon through the growth of farmers' capacity to interface with other supply chain agents over time and space. This will only occur if other producers develop distinctive approaches which are partly and variably associated with the use and identity of rural space and nature in new and innovative ways. In this sense we face one of the significant paradoxes of the new rural development paradigm. *How can clusters of rural development initiatives collectively make a major spatial impact when they are all – by definition – dependent upon a distinctive evolutionary trajectory?* Unlike the old CAP-driven agro-industrial system, there are few guarantees for producers. To succeed, participants in new short food supply chains need to rely heavily upon their own knowledge and networking abilities; as the Llyn case clearly testifies.

The development of short food supply chains also presents challenges to those institutions and agencies charged with supporting economic and social regeneration in rural areas. The opportunities and potential to develop local speciality products needs to be balanced against regional strategic considerations associated with more generic marketing and promotion. This is particularly significant where the value-adding potential of sub-regional identities lose their impact in export markets, and may hamper the ability to effectively supply and promote generic regional labels (such as Welsh Beef rather than Llyn Beef). The wider economic region (Wales) must effectively compete in major markets (exports to England and the rest of the EU), with quality products carrying appropriate branding, rather than having many sub-regional initiatives which may both intensify competition in domestic markets (bringing prices down at the farm gate) and simultaneously fail to engender the co-operation and co-ordination required to service major export markets.

Inside these sets of relationships identified here – and a major driving force behind their level of success – are the power struggles operating in the development of new food supply chains. Theoretically, we need to begin to assess how local ecologies, social relations, and forms of evaluation become implicated in differing production systems. Our analysis here demonstrates the significance of bringing together at least four types of evolution: temporal, spatial, demand and associational /institutional. From there we can start to specify what the most likely (RD potential generating) relationships between production process and place will be, and to question how the local social and natural resources of a particular place come to be incorporated into networks of food supply chains (formerly) dominated by the industrial and commercial modes of evaluation.

In particular, we can assess how struggles around new definitions of quality can empower local producers, such that they can develop sustained levels of value from production of food products. The development of new supply chain relationships is a key mode of rural development for achieving this; but it is one which suggests a heterogeneous and diverse set of rural development outcomes rather than a clearly generalisable model of convergent spatial development.

Notes

1 As a starting point here, Storper's (1997) definitions are adopted. These are ideal types of production linked to different types of market relationships (i.e. standardised, specialised, generic and dedicated; see Murdoch *et al.* (2000) for a more detailed commentary). While such categorisations are our descriptive starting point here, as we discuss further in the with reference to both the broader short supply chain analysis and the case study of the Llyn Beef Producers Co-operative, they are inadequate in capturing the evolutionary features of food supply chain construction. The analysis in this attempts to go beyond the rather static typological construction of supply chains, for it is only by doing this that we can begin to uncover the real social and material features that lead to the construction and development of short supply chains. Moreover, in conclusion, such explorations indicate the need for more dynamic conceptual formations which incorporate social uniqueness and difference into new rural development theory.

2 In the first phase of the 4th Framework FAIR research programme 'The socio-economic impact of rural development policies: realities and potentials' (IMPACT) (CT–4288) some 30 cases of rural development practices were studied. Of these 15 cases concern food supply chain initiatives, related to the fields of organic farming, high quality production, region-specific products and product diversification.

3 The case studies developed within the IMPACT project which have been analysed in the production of this are: (i) Llyn Beef Producers Co-operative – UK; (ii) Marches Quality Producers Group – UK; (iii) Graig Farm Organic Meat – UK; (iv) Organic Dairy Production in Wales – UK; (v) Farmers markets in the city of Frankfurt – Germany; (vi) Rhöngold dairy and organic farming – Germany; (vii) High quality food production in County Wexford – Ireland; (viii) Producer and consumer associations of ecological produce in Andalucía: an experience of short distribution channels of quality produce – Spain; (ix) From farmer to retailer: the last frontier for the raising of beef cattle in Umbria ? -Italy; (x) The production of Parmigiano-Reggiano cheese – Italy; (xi) Organic farming in Tuscany – Italy; (xii) Scalogno di Romagna – Italy,

4 In 1997, beef prices in Wales averaged 95.26p/kg compared to 121.55p/kg in 1995, a fall of 21.6 per cent (Table 10.1). Between 1996/97 and 1997/98, net farm income on lowland cattle and sheep farms in Wales fell by a dramatic 78 per cent from £13,101 to £2,914. As well as the BSE-induced crisis in confidence and the beef export ban, beef market realisations collapsed because of (a) a general trend away from consumption of red meat, (b) the appreciation of sterling (which significantly reduced the value of beef support schemes), (c) increased competition from imports, and (d) increased input costs over and above those 'normally' associated with beef production. These costs included the costs of establishing a database tracking cattle movement, and tagging.

11 Exploring empirical realities of rural development: the organic approach

Introduction

Achieving the triple dividend of (rural) economic development, environmental sustainability and the efficient production of safe, nutritious food is something of a Holy Grail for politicians, development agencies and many within the food industry itself – farmers included. The continuing 'race to the bottom' economics of primary production policy, is providing both a context for the agricultural crisis at the individual farm level (cost-price squeeze) and for broader issues and externalities against which the organic sector is able to differentiate and market itself (environmentally friendly, pesticide free, free of genetically modified organisms, etc.). Moreover, as we saw in the previous two chapters, new rural development policy options – designed to support the provision of environmental goods and services, and the marketing and processing of quality/speciality produce – are now supporting farmers who wish to shift their business emphasis away from commodity production and to preserve the European model of agriculture (Gray 2000).

The organic food sector appears to offer, in theory or rhetoric at least, some form of antidote to the concerns of society over food safety and environmental externalities, through a production system based on the sustainable use of natural resources and a rigorous system of certification. For some, the organic movement also has a political dimension, which may be understood as 'an oppositional politics of praxis that contests the industrial orderings of eco-social relations based on instrumentalist utilitarian rationality' (Goodman 1999, p. 32). Away from the more radical politics of sustainability advocated by some within the organic movement, organic farming has attracted much attention from policy-makers and politicians, many of whom are now subscribing to the rural development paradigm (Van der Ploeg *et al.* 2000).[1] Organics, as a form of high quality or differentiated production, is seen as having the capacity to extend new market opportunities and value-capturing rights, not to mention environmental benefits, down the food supply chain for the benefit of consumers, tax payers and local communities.

This chapter reflects upon the changing relationship between nature, society, political economy (of both production and consumption) and where the story of organics intersects with this, particularly as it relates to meeting new consumer needs, on the one hand, and providing a more legitimate and sustainable basis for public support to agriculture through the CAP on the other. It is argued that the development of the organic sector needs to be examined through such a prism in order to understand its competitive relations with the wider food sector and economy, which are now playing a central role in repositioning and reorganising organic foods and their supply chains. In so doing, I interpret the organic sector as part of the wider food sector rather than as something distinct – set up in oppositional or 'alternative' terms – and argue that as organics becomes subsumed into mainstream food supply networks several facets or elements of the more radical philosophy, particularly associated with local economic development and the support for small-scale agriculture, potentially become side-lined or ignored. Section two takes the case study of dairy production and processing to emphasise, in detail, the rural development impact of organics.

Finally, section three identifies how significant gaps exist between the realities of organics, when approached from a rural development perspective, and the model/theory of agro-ecology as elaborated by writers such as Altieri (1987) and Sevilla Guzmán and Woodgate (1997), which is identified as having a more holistic and socially oriented set of relations. In summary, and as illustrated in Figure 11.1, we believe that in order to understand relationship 'A' it is also necessary to understand relationship 'B', as one inevitably influences the context and potential for the other.

Figure 11.1 Interdependence and co-evolution between organic food supply chains and rural development potentials

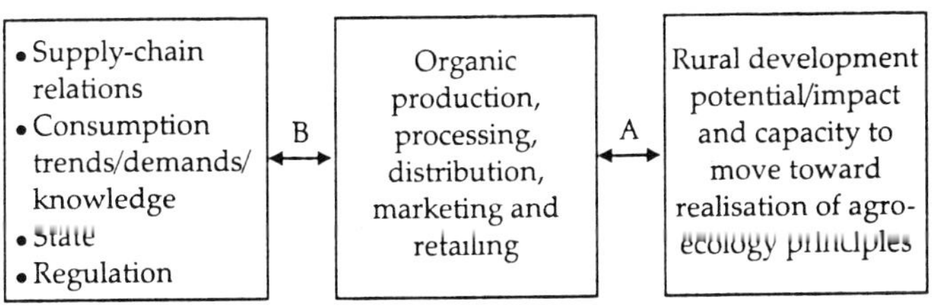

Perspectives on nature, society and organic foods: disconnections and distanciation

While the food supply chain is by no means the only vehicle through which society relates with nature and rural space (predominantly agricultural space in the UK), it does present a direct connection between nature, rural space and the life-styles and consumption choices of the general population. It is not surprising, however, that despite these connections, many citizens have scant knowledge of national or regional agriculture (NFU 2000). In developed countries, the vast majority of urban and rural populations alike now engage in economic activities unrelated to the land. Furthermore, consumption is disconnected in economic, social and spatial terms from places of production (Murdoch and Miele 1999). In what follows, I present different perspectives on the relationship between nature, society and economic relations as expressed through agro-food networks, and I use these to comment and reflect upon the growth and shaping of the emergent organic sector in the UK as it moves from the preserve of a specialised and relatively homogeneous (in ideological terms) group of producers and consumers to become what we might regard as a 'mass niche'.

Redefining and repositioning organics through the supply chain: production perspectives

The image of organic food producers and consumers, as well as the scale and professionalism of many organic food supply networks, have been transformed in recent years as demand for organically certified foods has accelerated (Buck *et al.* 1997; Tovey 1997; Banks 2000; Banks *et al.* 2000). Growth in the organic sector has resulted from both production and policy 'push' as well as consumer 'pull' influences. Without question, the provision of EU support under the agri-environmental regulation (Council Regulation (EEC) 2078/92, OJ no. L 215, 30.7.1992, pp. 85–90), has been a vital catalyst enabling farmers to seriously consider organic conversion.[2] In 2000, the organic food market in the UK was worth £ 546 m compared with only £ 200 m in 1997, and it accounted for about 1 per cent of the total amount spend on food (Soil Association 2000).

Prior to the mid 1990s, when it became evident that organics was more than just a passing fashion, and when both government and food retailers started to seriously engage with and address the opportunities presented by this sector, advocates and practitioners of organic food production were a very small and relatively homogenous group. This group pursued the organic system of production and consumption – often at significant opportunity costs to themselves as either farmers or consumers – in order to realise a personal belief in the holistic philosophy and practice of organic agriculture and living. While difficult to capture, this radical philosophy revolved around a belief in (i) the technical standards and practices of organic agriculture on farms linked to crop rotations and the minimal use of off-farm inputs and agro-chemicals; (ii) the practice of local production for local consumption – often involving interpersonal relations between growers and consumers (i.e. minimising food miles in the interests of broader sustainability) and support for small-scale agriculture; and (iii) a desire to consume foods that are not highly processed and that are also in season, thus reflecting a desire to work within rather than 'above' or in opposition to, natures cycles. In line with this model, committed consumers developed and used definitions of quality that challenged the highly sanitised and formalised quality conventions encouraged by the mainstream food supply sector, particularly in relation to shape, size, presence/absence of blemish (Murdoch *et al.* 2000).

As the market for organic foods has grown, the consumer base has broadened, and policy support has increased, the whole dynamic of the sector has changed. For example, research in Wales has shown that the days are now long gone when consumers or retailers were willing to pay premium prices for organic foods simply because they were certified (see Banks *et al.* 2000). In order to access many of the new markets (most notably supermarkets) and obtain viable returns, organic farmers must now meet stringent product specifications that reflect the quality conventions of the wider food system. Despite the requirement for organic producers to compete on the *quality treadmill* – a long established feature of more conventional food supply relations – organic farmers increasingly find their farms and their techniques the objects of neighbourly fascination.[3] A sea-change has occurred: rather than being an object of ridicule, organic farmers are now being characterised and perceived as progressives and innovators, rising to the challenges of emerging markets, and taking positive market-oriented steps to safeguard their futures in the face of increasingly volatile and price competitive commodity markets.

While difficult to substantiate, those now starting organic production from a background of conventional farming are more generally regarded as doing so in response to economic conditions. The degree of 'professionalism' and the eagerness of many new entrants (and some old hands) to capitalise on the opportunities presented by industrial /mainstream supply chains reflects the growing divergence in beliefs as to how the interests of the organic movement are best served (if they are even of concern). This raises an issue as to what extent organic producers, who are anonymous within their supply chains, will be subject to a greater cost-price squeeze effect when compared to those who remain more locally and regionally embedded, maintaining interpersonal relations with committed organic consumers. In the context of market contraction, those producers who are bound in close interpersonal networks of trust and respect, including common understandings and beliefs in a more radical holistic 'organic' agenda, may well find themselves better off, although the role of committed consumers who are willing to support local producers would be critical to such a scenario.

Marketing and retail

In the UK, about 80 per cent of grocery sales are conducted through the stores of 5 main retail multiples (Bell 1999). Consequently, food has become more anonymous, and consumers, unable to know the people (or even places) that produce foods, are now looking to the integrity of retailers, regulatory processes, and quality assurance schemes as a means of establishing or justifying trust in food. Moreover, such institutions and mechanisms are now pivotal in defining quality and value. In this context, 'organic' has become a powerful marketing device, drawing upon consumers' trust in sustainable farming techniques to deliver 'honest' and 'safe' food. Undoubtedly, the entry of multiples into the organic sector has served to develop the wider market for, and acceptance of, organic food. While the multiple retailers have only seen organic sales expand rapidly since 1997, many are now stocking upward of 450 organic product lines (the new norm among market leaders according to the Soil Association). The extent to which organics has taken off is exemplified by one high quality retailer, Waitrose, who reported that, in 2000, organic sales accounted for 12 per cent of all dairy sales, 10 per cent of all produce sales, 4 per cent of all grocery and 52 per cent of all baby-food sales. On a broader level, the Soil Association,[4] claims that 80 per cent of consumers have now consumed organic foods, although it should be stressed that the greater proportion of all organic food sales (63 per cent) are still accounted for by only a small proportion of consumers (7 per cent).

Table 11.1 Market share by final retailer, 1999

Point of sale	*Percentage of total sales (by value)*
Supermarket	74%
Independent retailers/health-food shops	17%
Farm gate/box schemes/market stalls	9%

Source: Soil Association (2000)

In much the same way as they dominate the conventional food retail sector, supermarkets are now the main outlet for organic foods (see Table 11.1). This contrasts with the situation a decade ago, when a far greater proportion of domestic and imported organic produce was distributed and retailed through box schemes, independent grocers or health-food shops. Unlike these more 'traditional' forms of organic retailing, supermarkets offer little to no scope for face-to-face contact or knowledge transfer between producers and consumers (other than that which can be squeezed onto packaging), and they can thus be seen as contributing to the growing gap of both knowledge and geography in food supply rather than reversing the trend.

While the entrance of supermarkets into the organic sector is welcomed by many for extending markets and the consumer base (thereby also providing more markets for farmers and leading to a greater area of organically managed land), there is growing evidence that this new form of competition is beginning to negatively impact upon existing, often more locally or regionally embedded, supply chains – most notably box schemes and producer cooperatives. This was recently exemplified by the voluntary liquidation of a major organic marketing company in England (OMC Ltd.), which had been a major supplier to box schemes and independent retailers. The company found itself unable to develop alternative markets quickly enough once the major supermarkets decided to give organic products serious attention, and more consumers who previously relied upon box schemes and 'alternative' forms of retail switched to making their organic purchases as part of a weekly supermarket shop. In light of this evidence, it is our view that while the system of organic production may still be regarded as 'alternative', it is now very difficult to argue that, for the majority of produce, systems of retailing and distribution can also defend a similar claim to 'alternative' status.

A consumer knowledge-based perspective

As the mainstream food industry has sought to engage with organics, emphasis appears to be shifting away from marketing and promotional activity directed at the socio-environmental benefits associated with organic production to a greater preoccupation with marketing, the claimed association between organic production, health and food safety.[5] With consumer confidence in the practice and regulation of conventional agriculture shaken by food scares and debates dating back over a decade (from salmonella in eggs in 1987, through to BSE in 1996, to genetically modified foods in 1999-2000), there can be little surprise that the organic label has been seized upon by consumers and retailers alike. Organics provides a clear basis for differentiating itself from other brands claiming to be 'natural' or 'environmentally friendly', as it is illegal for a product to call itself organic in the UK if it has not been certified by a government approved body.

Trust in, and knowledge of, labels such as 'organic' are critical to mass market appeal. Yet, as has already been noted, there is a growing chasm of geography and knowledge separating consumers from producers. Paradoxically, public awareness and knowledge linked to domestic agriculture, and even about how to handle and prepare food, appear to be diminishing as our general knowledge (or awareness) of the health and environmental impacts of certain foods and farming techniques are increasing. A recent survey by the National Farmers Union[6] showed, in their own words, that 'an alarming 70 per cent of the British public have no idea what food the farmers in their local area produce' (National Farmers Union press release, 1 June 2000). Further findings were that one third of people were unable to even hazard a guess at the main crops or food produced in their area, and that 70 per cent of those who did offer an answer were wrong. It concluded that 'the success of British farmers in producing a diverse range of produce from lavender to garlic and duck to sugar often goes unnoticed - even by the locals' (*ibid.*).

In parallel with mounting confusion over food issues, the increase in private disposable income means that more shoppers now have the opportunity to express their fears, aspirations or anxieties when they go shopping. Across a day, week or month, these may vary from being hedonistic, through individualistic, to outright altruistic. Product range diversification and value-added processing of organic products means that, as a sector, organics can now appeal to all three. Irrespective of motivation, ever more consumers are now able to connect with and understand organics, and this is proving a sufficient basis upon which to make consumption choices. This understanding is, however, constructed as much around what organic is *not* as around what it *is*, and grounded primarily on *how* rather than *where* or *who*. This was substantiated by a recent MORI poll published by the Soil Association (2000) which identified how one third of the public now buy organic food, primarily perceiving it as: healthy/better for you (53 per cent), tasting better (43 per cent), free of genetically modified organisms (30 per cent), and environmentally friendly/animal welfare friendly (25 per cent) (see also Table 11.2). Despite attempts to promote local food links and foreground wider connectivities between systems of food supply and sustainable development, the opportunity for close ties with producers does not appear to be highly significant in motivating consumption choices. Statements such as 'because it is local' fail to register or emerge from the MORI poll. The new consumer appears to be more selective and less holistic in defining sustainability, often focusing primarily on personal health-and-safety concerns rather than wider sustainability debates. Furthermore, food-miles concerns appear to be of no great significance to many new organic consumers, despite the obvious implications for the use of non-renewable energy and atmospheric pollution. The major retailers are also unlikely to risk limiting their opportunities by stimulating customers to dwell on these contradictions, especially in light of their own global sourcing strategies.

Kaltoft (1999) identifies how, within the organic sector, institutionalism involves a tendency to reduce diversity in practices and philosophies. Overviewing the sector, evidence suggests that organics is being increasingly subsumed into dominant forms of (centralised) distribution and retailing and that these processes are having major knock-on effects both on pre-existing supply-chain structures for wholesaling and retailing, on quality definitions and conventions and also on the homogeneity (in motivation and ideology) of producers and consumers. Through this process, organics risks being reduced from a distinct philosophy with profound implications for the way we produce, market and consume food, to just another form of product differentiation and category management.

Table 11.2 Key factors driving or motivating the growing consumption of organic foods

Driver	*Comment*
Consumer concerns over health	Consumers want to buy food perceived or designated as healthy Perceived potential health risk associated with the consumption of food additives used for preservation, colouring and flavouring within the food industry
Consumer concerns related to food safety	Concern over the health and environmental implications of genetically modified organisms Concern over potential danger of residues of agro-chemicals used in production of food under conventional systems
Environmental benefits	An increased awareness of the environmental damage associated with the use of modern agricultural techniques and the emergence of public dissatisfaction with the problems of European food surpluses
Food ethics	Concerns about animal welfare, local consumption and fair trade issues
Novelty and fashion	The desire for novelty and speciality in food consumption choices
Taste	Organic foods widely regarded by those who regularly consume them as having superior taste to conventionally produced alternatives
Access to organic food	The entry into organic retailing by the major retail multiples and the effect this is having upon access to organic food as well as competition for product within the supply chain

Source: Datamonitor (1999)

A political perspective

In parallel with snowballing successes in shops, organics has been widely embraced and promoted by politicians across Europe. The apparently ceaseless round of food health scares (often highly politically damaging and commonly tied to the regulation and practice of intensive production, or incompetence thereof), set against the back-cloth of mounting unrest over the budgetary and environmental impacts of productivist agricultural policy and the Common Agricultural Policy, have provided a framework through which the organic movement has been able to climb the political agenda.[7] Despite the association regularly made by consumers linking organic food with healthy food, the decision by the European Commission to support conversion to organic farming systems (and in some countries to offer continuing payments) continues to be based solely on a basis that the organic system of production yield environmental goods and services valued by the wider community and has nothing whatsoever to do with the relative merits of organic foods per se, as the following quote from the review of the Organic Aid Scheme makes clear:

> *The rationale for intervening to support organic farming will be to secure the socially optimal level of environmental goods. Supporting organic farming will not necessarily be the only or best way of securing those goods. However, to the extent that it is, it will be important to recall that the aim will not be to*

> *increase organic farming per se but to use organic farming as a mechanism to deliver an increased quantity of the environmental good required* (MAFF 1998).

This view is maintained because, despite all of the often evangelical claims made by companies trading organic products, the scientific jury is still out on whether organic farming always presents the most environment-friendly or safe-food option. MacKerron *et al.* (2000) and the Minister for Agriculture, among others, have rightly sought to stress that while there is nothing wrong with consumers demanding and paying well for organic foods, there is also nothing wrong with conventionally produced food, which is safe and often produced in highly 'sustainable' ways. They go on to argue that many conventional farms are managed in a way that delivers environmental benefits claimed by the organic movement, and that it is less intensive farming that delivers environmental returns not just certified organic practices alone. In this regard it is noteworthy that the Soil Association has recently had four judgements upheld against it by the Advertising Standards Authority in the UK. These claims included, 'You can taste the difference', 'It's healthy', 'It's better for the environment' and 'Organic means healthy happy animals' (ASA 2000).

The decision by government to increase the levels of funding to support organic production, and in Wales the decision by development agencies and public–private partnerships in the agro-food sector to develop and promote the organic sector,[8] reflects how the institutions and politicians are now acknowledging the role and potential of organic food supply to meet new societal demands in the shops, social and political aspirations relating to rural environments, as well as to offer new market opportunities for struggling farmers. The next section of the chapter will continue to explore this rural development side of the organic equation through reference to the case study of organic dairy production and processing in Wales.

Does the organic equation add up? The rural development question

The potential for food supply chains, particularly high value-adding 'alternative' or 'short food supply chains', to act as rural development mechanisms has been addressed in recent academic debate within rural sociology (see Banks and Bristow 1999; Marsden *et al.* 2000a). The argument is that if consumers are willing to pay a premium (higher) price for products with a certain provenance, quality certification or deriving from a specific type of production (such as artisan or certified 'organic'), opportunities exist for farmers to capture a greater proportion of the total value added from farm to fork. These opportunities, it is further argued, can be increased if farmers are producing scarce commodities, and even more so if they are able to either increase value added themselves through processing or become involved further down the supply chain, such as in direct sales at farmers' markets.

The case study of organic dairy production and processing in Wales presented below was drawn up as part of the IMPACT programme in 1999. It seeks to exemplify how rural development gains have been generated by the growing market for organic foods, and why the option of conversion to organic status is now being regarded as one key trajectory for agricultural development and the agro-food sector in Wales. The case study focuses on how rural development is realised, with a particular focus that is placed upon the role of organic supply chains in removing or mitigating the cost-price squeeze being experienced by conventional producers.

Case study: organic milk production and processing in Wales

Organic dairy production has a long history in Wales. The first Welsh organically certified farm was established by the Rowlands family back in the 1940s. This family would subsequently go on to establish the largest organic dairy in Wales, trading under the name 'Rachel's Dairy' – now owned by the largest organic dairy processing company in the USA (Horizon Organic Dairy) following a sell-out in 1999. Despite product innovation, market success, and an expansion in the processing capacity of Rachel's Dairy in the early 1990s, relatively few farmers were interested in producing organic milk in Wales until after the review of the Organic Aid Scheme in 1998 which increased the payments to support conversion. Furthermore, the conventional milk market was delivering good returns (€ 0.4 per litre) prior to 1997, and because the organic market was relatively embryonic, the price premium to organic producers was only small (€ 0.05 per litre). At this time those practicing organic milk production were doing so largely due to a commitment to the principles of organic farming or in association with speciality on-farm processing. As an indicator of growth, we estimate that in 1998 there were only 11 fully certified organic milk producers in Wales producing around 5 million litres of milk a year (Banks 2000), but that by 2001 there were an estimated 55 fully certified organic dairy farms in Wales with a further 90 in conversion.[9]

Table 11.3 illustrates the growing interest in organic conversion by showing the number of inquiries made to the Organic Conversion Information Service (OCIS)[10] and the number of farm visits undertaken, both to all farm types and separately to dairy farms (also given as a percentage of all visits).

Table 11.3 The Organic Conversion Information Service in Wales, 1997–2000

	1997	*1998*	*1999*	*2000*
Total no. enquiries to OCIS	293	756	939	452
Total no. of farm visits	196	415	541	300
Total visits to dairy farms	60 (31%)	84 (20%)	103 (19%)	69 (23%)

Source: Soil Association (2001)

Growing interest in organic dairy production has mirrored the emergence of a substantial price differential between standard and organic milk. In recognition of this divergence, organic milk producers succeeded in decoupling the organic milk price from the conventional milk price in 1998, that is, the organic price was no longer calculated from a premium on top of the conventional milk price, which was falling. Table 11.4 gives details of how prices have changed since the mid-1990s for both organic and conventional milk. It should be noted that the Organic Milk Supplies Cooperative (OMSCO) was not operating in Wales until 1999, but the prices paid to organic milk producers in Wales (most of whom traded through the Milk Marque cooperative) were broadly in line with those of OMSCO prior to this time. For the period covered by the statistics, the majority of farmers in Wales supplied milk to processing companies through the Milk Marque cooperative, which also set the basic farm-gate price for milk nationally; however, the prices of other conventional milk buyers are also included.

Despite the emergence of a price differential worth over € 0.17 per litre evidence from a survey in Wales suggests that farmers considered conversion to be too difficult to pursue purely for economic reasons. In addition to the obvious market opportunities presented by shifting away from undifferentiated commodity production, environmental benefits and animal welfare implications of organic production were both sited alongside the fact that for some farms the conversion process would not represent a major change in their farming practices (see Banks *et al.* 2000)

The 'value added' of organic production on farms

Gross margin calculations can be used, with some degree of caution (see Lampkin and Measures 2000) to compare the costs and incomes derived from organic and conventional production systems. Table 11.5 brings together comparable statistics on the costings of organic and non-organic diary herds.

Table 11.4 Extract of the producer price league table

	Price per litre (Euro)			*Premiums over Milk Marque (Euro/litre)*		
	1995/1996[a]	*1996/1997*[b]	*1998*[c]	*1995/1996*	*1996/1997*	1998
Organic Milk Suppliers Co-operative	0.451	0.470	0.479	0.054	0.114	0.179
Golden Vale (non-organic)	0.420	0.388	0.336	0.023	0.032	0.036
Associated Co-operative Creameries (non-organic)	0.411	0.388	0.318	0.018	0.032	0.019
Unigate (non-organic)	0.412	0.376	0.320	0.015	0.020	0.020
Milk Marque (non organic)	0.397	0.356	0.299	0.000	0.000	0.000

Source: Competition Commission (2000)

Notes:

[a] Net of transport and all deductions for milk with 4.1 per cent butterfat and 3.25 per cent protein, with somatic cell counts of 200,000/ml and total bacterial counts of 10,000/ml or Bactoscans of under 100. Ranked on rolling 12 month cumulative prices actually paid.

[b] Rolling 12 month cumulative prices actually paid for October 1996–September 1997, net of transport and all deductions for milk with 4.1 per cent butterfat and 3.25 per cent protein, with somatic cell counts of 150,000/ml and total bacterial counts of 10,000/ml or Bactoscans of 50.

[c] Milk prices and bonuses paid for 12 month period from January 1998 to December 1998, for milk supplied with 4.1 per cent butterfat and 3.2 per cent protein with somatic cell counts of 150,000/ml and total bacterial counts of 10,000/ml, sending a volume of 600,000 litres per year.

Table 11.5 Conventional and organic milk production variables, 1999, in Euros

	Conventionally produced milk	*Organic milk production*
Milk value per cow (per year)	*1799.82*	2677.86
Variable costs (per cow per year)	*528.12*	612.36
Gross margin per cow	*1088.64*	1654.02
Gross margin per forage hectare	*2178.9*	3141.18

Source: Nix (2000); Lampkin and Measures (2000)

The € 565 difference in margin per cow between the two different systems equates to a total income differential of € 33,923 for herds of 60 cows over one year. However, in undertaking this calculation we must be mindful that conventional systems are likely to contain a greater number of stock. Nix (2000), in his calculations of conventional dairying, assumes an output of 6350 litres per cow per year and a stocking density of 2 livestock units per hectare with a milk price of € 0.28 per litre after deducting transport costs. Lampkin and Measures (2000) estimate on only 5,700 litres per cow and base their estimates on a stocking density of 1.7 livestock units per ha, with the milk price being € 0.47 per litre. Further calculations based upon information gathered from farmers in Wales broadly align with these results. Table 11.6 clearly shows the financial benefits associated with organic production at current market prices and is designed to reflect more closely the situation in Wales in 1999. Even allowing for the fact that conventional herds will be 20 per cent larger and produce over 1,000 litres per year more per cow, the difference in income between conventional and organic systems is still € 2,648 in favour of the organic system on a herd of 55 cows when compared to a conventional herd (average performance) of 66 cows.

Table 11.6 Organic versus conventional milk production in Wales, 1998–1999

	1999 conventional at € 0.29/litre[a]	1998–1999 organic at € 0.47/litre[b]
Total output		
€ / cow	1,743	2,297
€ / litre	0.29	0.46
Total variable costs		
€ / cow	744	612
€ / litre	0.12	0.12
Gross margin incl. Forage		
€/cow	1,003	1,685
€ / litre	0.17	0.34
Margin for a herd of 55 (organic) and 66 (conventional)[c]	66,183	92,664

Source: Banks (forthcoming)

Notes:

[a] Calculations based on 6,091 litres per cow per year; milk price of £ 0.18 per litre. Slight reduction in variable costs has been made to account for greater cost awareness as milk prices have fallen (by 20 per cent).

[b] Calculations based on yield of 5,000 litres per cow per year. No reduction in variable costs has been made.

[c] An increased herd size of 20 per cent has been allowed for conventional units to take into consideration the higher stocking densities possible in non-organic systems.

At the farm level, it can thus be seen that organic milk production does currently constitute one potential answer to the cost-price squeeze experienced by productivist agriculture, by enabling the farmer to acquire new financial benefits and/or realise new forms of cost reduction resulting in an increased economic margin/value added. Of course, it is likely that in the future the prices paid for organic milk will reduce, as would be expected when supply comes into line with demand. Through effective cooperation and contract negotiation, however, farmers have the opportunity to safeguard their incomes, particularly in view of the anticipated growth in the market and dairy processors' desires to secure volumes for value-adding processing.

Supply-chain relations and their capacity to impact upon rural development potentialities

The rapid rate of farm conversions has resulted in Wales switching from being an organic milk importer to a milk exporting region. Two further factors underpin this change and have implications for the ability of Wales to maximise the potential rural development impact of the organic dairy sector. First, as of 2001 there were four organic milk marketing cooperatives operating in Wales, three of which supply Rachel's Dairy, whose main outputs are fresh dairy products such as yoghurt, cream and soft cheeses. All of the groups claim to have growing memberships and volumes under their control.[12] Unfortunately for suppliers, whereas it may take 10,000 litres of milk to produce a ton of cheese, this ratio is much reduced in the manufacture of organic yoghurt and other fresh products. The capacity of the major processor to utilise raw milk is therefore currently limited. This brings us to the second point, the fragmented nature of the supply base means that other processors who could provide alternative markets within Wales are finding it very difficult to coordinate and negotiate supplies from these groups to establish product lines, particularly as significant volumes are required to undertake production runs in larger dairies. It is necessary for a large plant/company to enter into organic processing to undertake product development, market research, produce packaging designs and organise distribution and marketing activities. Such activities are costly and, if they are to be undertaken, require a company to have secured access to a regular supply of milk so that costs can be recouped through sales. For a large dairy it may take 50,000 litres of milk a day to start a production line and, at present, this would represent the milk from about 75 per cent of currently existing organic dairy farms. Hence, until more producers emerge from conversion, these volumes are unavailable on a regular basis. Some smaller processors, able to handle smaller volumes, have started to utilise some of the excess capacity, but even here difficulties exist.

Through more effective coordination within the supply chain, producers could be helping regionally based companies develop products and capture market share, rather than passing these opportunities to companies and farmers outside of Wales. Product development and investments in processing capacity have a vital role to play maximising the overall returns from the organic food sector. This understanding is reflected clearly in the strategic aims of the Organic Sector Action Plan drawn up by a public–private partnership in Wales in 1999.[13] The ability to retain value-adding capacity within rural Wales is highly significant in expanding the overall rural development impact of the organic sector.

The wider impacts of organic dairy production in Wales: economic multipliers

Statistical analysis of economic multipliers now exists for various key food sectors in Wales, including the dairy processing sector (Hill and

Roberts 1998). From this information it is possible to gain insights into the relative importance of the dairy sector in the Welsh economy, as well as the knock-on impacts of investment and employment generation. For the purpose of this brief analysis, it is assumed that the impact of the organic processing sector will be equal to, but not greater than, that of the dairy sector as a whole.

Input–output tables can be used to assess the economic impact of various changes, such as the opening of a new factory, on the Welsh economy. The analysis is conducted in terms of economic multipliers that enable the estimation of both direct and indirect or spillover effects of change to be predicted. The spillover effects are twofold. First, they occur as a result of local supplier effects, whereby an expanding industry sector requires more supplies, some of which will be bought locally, thereby generating demand for the outputs of local firms. Second, there will be an income-induced effect, whereby an increase in demand for an industry's outputs will mean that the industry requires more labour. This in turn, will add to local income from employment, some of which will again be spent locally, thus generating further demand for local industrial output. Supplier effects (i.e. non-local sourcing effects) are conventionally termed 'type I multipliers', whilst the total multiplier, including induced-income effects, are called 'type II multipliers'. From the figures presented in Table 11.7 we can thus estimate that if 100 full-time employment equivalents (FTEs) are created in the dairy processing sector, these will result in an additional 179 FTEs through supplier linkages (type I multiplier) and an additional 219 FTEs when the spend of those original 279 FTEs (179 net) is added (type II multiplier). Similarly, it can also be calculated that an increase in demand for organic products of € 100 will result in € 142 of Welsh output, including supplier effects, and that when the induced-income effects are added, the initial € 100 ultimately generates € 151 of output in Wales.

Table 11.7 Inputs, outputs and economic multipliers for the dairy sector in Wales

	Dairy processing sector
Total value added	€ 283.7 m
Employment (full-time employment equivalent)	2577
Value added per full-time employment equivalent	€ 110,074
Output type I multipliers (direct)	1.424
Output type II multipliers (direct & indirect)	1.509
Employment type I multipliers (direct)	2.789
Employment type II multipliers (direct & indirect)	3.194

Source: Hill and Roberts (1998)

Returning to the organic sector in Wales a strong case can be made that all the employment generated and investment resulting from the Rachel's Dairy operation can be considered a rural development gain resulting from the existence of both a market for organic products and the availability of organic milk for processing. Estimating the employment of the company to be around 40 FTEs (1999), it can thus be calculated that the business supports an additional 72 FTE through related supplier linkages, and that, if the spend of those extra employees are taken into account, the impact on the wider economy (which will not all be accounted for by expenditure in the rural economy) will be to generate 87 FTEs over and above the 40 FTE already engaged at the factory itself.

Concluding comments

For those with the motivation and skills to operate successful diary farms, current market conditions provide opportunities for significant value-added benefits over conventional dairy production. Farms can supply existing buyers or attempt to develop on-farm processing if they wish to extend their operations down the supply chain. However, the opportunities presented by the organic markets should not be regarded as a panacea for all struggling dairy farmers. The existing structure of subsidies and support payments (often headage based) mitigate the successful entry and participation of many very small or upland/hill-based dairy farmers, most notably where conversion necessitates a reduction in stocking densities. Furthermore, it cannot go without mention that the viability of farms will depend on other activities and crops/outputs, and the ability to control costs through effective management practice. Farmers undergoing conversion also need to be prepared for market conditions to evolve. It is likely that as more farms become certified and the scale of production increases, producers will be expected to meet more stringent quality standards for organic milk, and that pressures to reduce prices will intensify. The strategic activities of farmers in particular may have major implications both on their ability to maintain returns at current levels, and also on the further development of processing in Wales. With this in mind it is essential that growth in supply is paralleled by expansion in processing capacity.

Re-appraising organics and rural development through principles of agro-ecology

The previous section has demonstrated how, in principle, organic production and processing can delivery rural development gains at both the farm level and within regional economies through processing activities and supplier linkages. However, the focus of the analysis is unashamedly economic. In order to counter this bias, this final section takes the study of organic food supply chains and their development in the UK – notably the example of organic dairying – and assesses the extent to which these experiences correspond to the principles of agro-ecology as elaborated by Sevilla Guzmán and Woodgate (1999) and outlined in earlier chapters in this book. This approach is of use in highlighting the divergence and distance between the supply chain and consumption relations of organics in the UK at present and the more holistic and socially oriented relations advocated by the agro-ecology model.

The principles and definitions of agro-ecology

In many developing countries (and some parts of the developed world) agro-ecology as both a social and political movement and a set of principles for sustainable food production is gaining momentum (see Parrott and Marsden 2001). As a concept, agro-ecology is a broad church, but at its heart lies a desire to practice environmentally and socially sensitive approaches to agriculture. Altieri, a leading proponent, sees agro-ecology as an 'alternative agriculture', stating that,

> *'Alternative Agriculture' is defined here as any approach to farming that attempts to provide sustainable yields through the use of ecologically sound management technologies. Strategies rely on ecological concepts, such that management results in optimum recycling of nutrients and organic matter, closed energy flows, balanced pest populations and enhanced multiple use of landscape* (Altieri 1987, p. xiv).

Hence, agro-ecology shares much common ground with the principles of organic farming, notably the desire to promote and undertake environmentally sustainable forms of production. Extending the definition, Hecht explains how agro-ecology also encompasses a broader remit. She observes,

> *The term agro-ecology has come to mean many things. Loosely defined, agro-ecology often incorporates ideas about a more environmentally and socially sensitive approach to agriculture, one that focuses not only on production, but also on the ecological sustainability of the production system, This might be called the 'normative' or 'prescriptive' use of the term agro-ecology, because it implies a number of features about society and production that go well beyond the limits of the agricultural field* (Hecht 1987, p. 4).

Elaborating this broader agenda and seeking to establish a set of agro-ecology principles, Sevilla Guzmán and Woodgate identify agro-ecology as promoting the,

> *eco-management of biological systems through collective forms of social action, which redirect the course of co-evolution between nature and society in order to address 'the crisis of modernity'. This is to be achieved by systemic strategies that control the development of the forces and relations of production in order selectively to change modes of human production and consumption that have changed this crisis. Central to such strategies is the local dimension where we encounter endogenous potential encoded within knowledge systems (local, peasant or indigenous) that demonstrate and promote both ecological and*

cultural diversity. Such diversity should form the basis of dynamic yet sustainable rural societies (Sevilla Guzmán and Woodgate 1999, p. 83).

This definition emphasises the impact that the realisation of agro-ecology principles is anticipated to have on the organisation of social and economic relations, and appears to have as much to do with maintaining viability for rural communities, especially farming communities and cultures, as with specific forms of production. The statement contains many ideas that supporters and practitioners of a more integrated and holistic organic/sustainability philosophy would be sympathetic toward. These include the desire to reassert localism, to embrace endogenous skills and knowledge, to establish or assert new norms of production and consumption practices, and to value diversity in both the scale and practice of farming. For, as Goodman observes:

> *In organic agriculture the fetishised abstraction of food is intentionally unveiled, bringing the complex filaments of food provisioning explicitly into focus. That is, the organic agro-food network invites scrutiny of its constituent metabolic relations, an interrogation that follows from its organisational and ethical premises of connectivity, in contrast to the punctualisation of black boxing characteristics of industrial agro-food networks. The multiple human and non-human actants involved at every step are explicitly foregrounded, giving transparency to the complex material, eco-social, and discursive practices that bring food from the land to the table in the routine context of everyday life. In these alternative actor-networks, food becomes a signifier for political, social and ecological struggles that are otherwise easily ignored* (Goodman 1999, p. 32).

UK organics in relation to the principles of agro-ecology

The principles of agro-ecology as set out by Sevilla Guzmán and Woodgate (1999) and the idealised model of organic agriculture as expressed by Goodman present a vision, but one that is complicated by the complex interaction of production with supply-chain relations, methods of distribution and retail, and the motivations and understandings of consumers. In the following, we use the example of organic dairy production and processing in Wales, alongside some broader references to organics in the UK, to briefly assess the degree to which the organic sector is leading to, or reflecting, a fulfilment of some of these principles. Some of the parameters of agro-ecology as established and expanded upon in the introduction to this volume (see Marsden *et al.* 2001a) are used as a point of departure.

Crisis of modernity

'Agro-ecology implies an alternative definition of sustainability from which is generated on ecologically – rather than industrially oriented discourse.' Despite advocating a very different set of practices on the farm, evidence from the dairy sector (and others) in Wales suggests that in many ways the dimensions and discourses of organic food supply do little to challenge or provide an alternative to 'industrial' discourses through the wider supply structure. In dairying, for instance, the milk price structures are designed around the production of milk with specific qualities (i.e. constituent values of butterfat and protein linked to processing requirements, as well as hygienic qualities that, while important for consumer safety, have relevance for product shelf life and hence 'durability over distance' (see Coombes and Campbell 1998). Moreover, energy-intensive forms of packaging and high food miles, linked to the servicing of national markets, do nothing to challenge the mainstream process of marketing and distribution. The target market for the majority of organic foods now produced in Wales are the retail multiples (although other significant markets do exist, many more local or regional in focus). Such markets are very much part of the identified 'industrial' discourse of agro-food supply.

Local farmers' knowledge systems

'The local dimension is central to strategy where we encounter endogenous potential that we find encoded in knowledge systems that allow for and promote both ecological and cultural diversity.' The organic farm owned and operated by the Rowland's family in Wales was a pioneering unit being certified in the 1940s. Consequently, the farm has never been managed in accordance with the principles of modernisation so vigorously encouraged in the post-war period and almost universally adopted elsewhere in Wales. However, to claim that the more widespread adoption of organic farming techniques is based upon 'rediscovered' or 'revalued' forms of local knowledge is, we believe, false. Those farms now adopting organic techniques are primarily doing so for the first time, having spent a lifetime applying the techniques and management practices of intensive agriculture in the pursuit of increased yields and specialisation. The new converts to organics now reversing their intensive practices are doing so not on the basis of local or traditional knowledge networks, but in accordance and under the strict guidance of national organic standards and the extension support network that accompanies it. What we might understand as 'endogenous' or 'local' knowledge today equates to efficiency-oriented, 'industrial mode' agricultural techniques as those currently farming have always operated within frameworks and systems that do not challenge the long-established modernist paradigm – now over 50 years old in the UK.

Collective forms of socialisation

Discussions in the first section of this chapter have sought to demonstrate how, while the organic sector continues to offer opportunities for producers and consumers to engage in 'collective forms of social action' such as box schemes and other cooperatives, the competitive environment in which such activities are reproduced has been transformed by the concerted entry into organic retailing (and food processing activities via own-brand labels) of the major supermarkets. While 'agro-ecology thus is founded upon new forms of associationalism' (Marsden *et al.* 2001a), the evidence suggests that organic sector growth has not encouraged and supported the expansion of those alternative networks and associations that do exist – more likely they are struggling to compete against mainstream alternatives now also offering organic lines.

Global reach and international sourcing are now as much a part of the organic sector as they are of conventionally produced foods. The proportion of organic foods being traded through locally supplied box schemes is now accounting for a far smaller proportion of total organic sales than was the case in the early 1990s, primarily due to the fact that more consumers are now able (and prepared) to make their organic purchases in the supermarket. The entry of major supermarkets and other multinationals into the organic movement does not preclude the 'eco-management of biological resources through collective forms of social action' at the local level, as idealised by Sevilla Guzmán and Woodgate (1999). However, the physical and intellectual distantiation engendered by such large-scale corporate forms of food distribution and retailing can mitigate against this principle being realised. Furthermore, there can be little true justification for claiming a fundamental change in the 'mode of human production and consumption' when the organic industry (often mainstream multinationals trading under less obvious titles or offering retailer 'own-brand labels') is offering a menu of high-processed and heavily packaged foods, the ingredients of which may have come from the four corners of the globe. This raises the question of whether organics as a countermovement and radical 'holistic' philosophy of sustainability (paralleling agro-ecology principles) can really survive in a global trading environment? Moreover, it raises a challenge to the organic movement to raise the awareness and value placed on food supply by new consumers of ecological and social discourses (in addition to health and safety-related discourses) and make these equally important motivations for consumption choices.

Sustainable societies

The case of organic dairy production in Wales indicates that this activity is doing little to reduce the segregation of agro-food production from rural society - of which a definition is extremely difficult to determine. Sevilla Guzmán and Woodgate, however, argue:

> *Sustainable societies can only be constructed on the basis of sustainable, locally relevant agricultures ... implying a complete rejection of the homogenising tendencies of the neo-liberal, global modernisation project and the re-direction*

> *of co-evolution towards more sustainable ways of living that are based upon the endogenous potential of an infinite diversity of locally relevant agro-ecosystems* (1999, p. 304).

While it is perhaps unfair to expect the organic movement and the operation or organic food supply to spearhead such a major political, social and economic project, it must still be observed that 70 per cent of the organic foods consumed in the UK are imported from overseas. Closer to home, and with reference to attempts to transform 'co-evolution', we observe that the employment gains resulting from organic farming are only small and are likely to have a negligible impacts when agricultural employment constitutes only a very small proportion of rural employment. More positively, however, organic production does emphasise the multifunctionality of farming systems, in particular their ability and role in the provision of environmental goods and services. Uniquely in Europe, support to organic farmers via the Organic Farming Scheme only covers the period during conversion, the market then providing what are perceived as adequate returns to sustain such practices (and thereby removing the need of the state to intervene as a result of 'market failure' to support the provision of environmental goods). While this may be interpreted as consumers demonstrating a willingness to pay for such environmental benefits directly, evidence from consumer surveys reveals that other factors (notably health and genetically modified organisms-free status) are key determinants to many; this raises the question as to what extent a shift in co-evolution is vulnerable if not based upon broader socio-ecological goals?

Principles, models and realities

Until the 1990s, the ethos of most businesses within the organic sector, and the motivations of organic consumers, were arguably more homogenous than they are today. A greater proportion of consumers had an understanding of the 'alternative' organic philosophy, and were prepared to support forms of distribution and retailing that reasserted localism and interpersonal communication within the supply chain. The desire to radically break away from the conventions of the mainstream agro-food sector doubtless contributed to those in the organic movement being regarded as 'subversives'. However, as already alluded to, the upscaling of organic processing activities and the entry of major supermarkets into the organic sector, and the health and food safety motivations of a majority of consumers now heralded a new era for the organic movement. Resultingly, the outcome appears to be a divergence in realisation and ideology between the principles of agro-ecology and the realities of organic production and consumption to the point where one is no longer clearly leading toward the realisation of the other.

What emerges from this discussion is a realisation that the growth in *the organic sector does not necessarily lead directly to the realisation of the principles of agro-ecology – or even the principles of the radical organic movement itself.* Put another way, it is not the case that any policy for generating organic agriculture is necessarily going to lead the way to agro-ecology benefits. We, therefore, need to look analytically not only at the rural development impacts emerging from the practice of organic agriculture and product transformation and marketing, but also at the particular *supply-chain relations and agents to understand which supply chains empower which actors over others,* and the degree to which asymmetric relations can be challenged and ecologically oriented discourses foregrounded. By the same token, we believe that if agro-ecology is to become a meaningful discourse in the UK context, it must embrace an analysis of farms' interactions and engagements with food supply chains and the broader context of capitalist reproduction and regulation, as activities downstream from the farm now play a key role in defining the conditions for, and definition of, organic and sustainable forms of production.

In summary, we can see real difficulties in shifting organics into the principles of agro-ecology, and a dual process of institutionalisation is occurring that mitigates against this realisation. The first difficulty is that which is resulting from interaction and engagement with (or subsumption into) the mainstream food supply sector. This we can see as leading to an emphasis in supply relations and marketing activities, which often obscure wider social and ecological dimensions related to what Goodman refers to as the 'organisational and ethical premises of connectivity' (Goodman 1999, p. 32). The second form of institutionalisation is associated with the process of formalisation/legalisation in relation to technical standards – if you like, the bureaucratisation of organics. Through this, we can see organics becoming more legalistic and bureaucratic, with a focus upon adherence to identified technical procedures and practices irrespective of the broader sustainable or unsustainable context in which they are undertaken.

Conclusion: dealing with the contradictions

The organic movement has always faced a dilemma. On the one hand, it has wished to expand the market for organic foods leading to more organic land and a greater number of consumers. However, the most obvious way of achieving this – that is, through engaging with supermarkets who act as a key gateway to the mass market – has meant engaging with conventional supply chains, and also expanding the opportunities for those perhaps less committed to a radical philosophy of sustainable development. The opportunities for profit available in the expanding organic market are now leading to a potential side-lining of the social and local agendas. Those interested in promoting organic as an end in itself, trading on the environmental and healthy associations cached to organic foods, also run the risk of compromising the very integrity of that which they seek to capitalise upon, particularly when offering goods that, because they have been highly processed or transported over vast distances (with hidden environmental costs), fail in relation to a holistic understanding of organics, even if they meet the technical standards.

Speaking at a recent conference on organic food in the UK, the director of the Soil Association stated that, '[u]nless there is education into organics, much of the potential wider benefits for society and the development of local economies will be lost. It would be a tragedy' (Patrick Holden, Cirencester 2001).

This raises the question of organics development in the UK at present, either as a type of product differentiation and/or as a broader socio-political and consumer movement. One issue appears to be that, in order to extend the rural development and eco-social sustainability impacts of organics, we need to initiate new forms of institutional support. Those currently in operation are doing little to fundamentally shift the pattern of co-evolution of nature and society. In other words, we need to change more than just the way we go about the socio-technical process of producing organic food (of which organics is a clearly identifiable alternative to currently unsustainable practices) but also to change the approach of the state to agricultural support, consumption patterns (deriving from educated and interested consumers), retail structures, processing and supply-chain relationships. In view of this, organic development as it is currently being expressed in the UK context is having a so far limited role and impact in line with the precepts of agro-ecology, even if it is able to deliver some much needed rural development impacts. The principles of agro-ecology places at their centre social components, rather than capital; people within production systems and people in communities become the focus rather than the narrow notions of 'efficiency' of production processes themselves. At present, for the majority of consumers organic food the priority is with quality and assurance. It is not seen as an issue for maintaining viable incomes for small-scale producers or battling against globalised agri-business. Hence, we see here, as argued in Chapter 8, even within a clearly defined 'agro-ecological' sector – like organics – there is a persistent tendency to *fracture and distance the links between sustainable production, rural development, healthy food consumption*. This fracturing process provides real obstacles for the capacity of the organic movement to be a major countermovement away from 'race to the bottom' agro-industrial agriculture and towards the new rural development paradigm. As markets have expanded, the consumer base has seemingly become less homogenous and less educated as to the philosophies and eco-social principles upon which organics as a political movement is grounded. It, therefore, appears that as social sustainability becomes the misunderstood relation in the sustainability equation of organic farming, the rural development potential of the activity may paradoxically diminish.

Notes

1 The rural development paradigm seeks to move beyond the post-productivist countryside (Marsden *et al.* 1993) in recognising that agriculture and food supply chains can play a core function in delivering sustainable development through strategies such as quality production, production and place differentiation, adding-value (as opposed to commodity production) and new forms of marketing.

2 The costs of conversion can be high for many farms, involving a reduction in yields, and the process may take several years. During the conversion period, farms must continue to sell their output through non-organic supply chains, where the reduced volumes and potentially lower specifications can result in a steep drop in income until the 'premium' organic market can be accessed upon certification.

3 CAP production subsidies are paid in Euro and not sterling at the national level. When the sterling appreciates in value relative to the Euro, the impact is that the subsidy payments translate to fewer sterling, therefore reducing the income of farms. During the late 1990s and 2000, the strength of sterling was seen as underpinning the farm income crisis.

4 The Soil Association has been the main advocate of organic farming in the UK. Founded in 1946, members shared a concern about the health implications of increasingly intensive agricultural systems being developed at that time. The principle concerns were (i) the loss of soil through erosion and depletion, (ii) decreased nutritional quality of intensively produced food, (iii) exploitation of animals in intensive units, and (iv) impact of large intensive farming systems on the countryside and wildlife. While the issue of local production for local consumption and maintaining the viability of small farms and rural economies were not explicit aims of the Soil Association, they have subsequently been developed through the practice of encouraging local 'food links'.

5 Tesco adverts run in 2000 focused on unsubstantiated claims that organics is a 'healthy' food, rather than the more substantial claims that can be made about the environmental impact of organic food production techniques.

6 Research conducted for the National Farmers Union by Taylor Nelson Sofres (National Farmers Union press release, 1 June 2000).

7 For example, intensive agriculture has caused farm-land bird populations to decline faster in the UK than in any other European country, with the Royal Society for the Protection of Birds (a potentially significant lobby group with over 1 million members nationally) producing data that claim to show a direct link between population declines in farm-land birds and indicators of agricultural intensification, including yields and tractor numbers (Farmers Weekly International News, 7 January 2001).

8 In March 1999, representatives from both public and private sectors (farmers, processors, retailers, policy-makers, development agencies etc.) came together to develop strategic action plans for the development of the organic sector in Wales – recognising that opportunities existed. The outcome was an Action Plan that included plans for the formation of an Organic Centre of Excellence to bring together expertise and advice and provide a 'one-stop-shop' for organic farmers.

9 Estimates based on interviews with WDA Organic Marketing Officer, for 2000, with data from Banks (2000) for 1997.

10 In 2000, Milk Marque was disbanded following an inquiry by the Competition Agency and reformed as three smaller regional producer cooperatives. Most of the farmers in Wales, formerly holding contracts with Milk Marque, now supply milk through a company called Axis.

11 OCIS operates nationally and is the first point of contact for farmers wishing to consider organic conversion. The service provides a half-day visit by an officer who takes an interested farm through what organic conversion might mean for the particular farm and to answer any questions. A second day-long visit is then arranged during which a farm conversion plan may be drawn up prior to submitting an application to the Organic Aid Scheme.

12 The organic milk groups currently operating in Wales are: (i) Axis, (ii) OMSCO, (iii) Calon Wen, (iv) Zenith.

13 See note 8.

12 Conclusions: rural development as 'real' ecological modernisation?

Introduction

Section 3 of the book has taken a markedly different perspective in examining the character of rural development. It has placed emphasis upon the ways in which rural and agricultural 'natures' *may* be more effectively incorporated into our understanding of rural change, and, particularly how this might lead to more sustainable rural development in the future. The exploration has attempted to bring together an appreciation of rural development practices and prospects on the one hand with theories of agro-ecology and ecological modernisation on the other. There has been a tendency for the latter bodies of theory to be far too restrictive, with agro-ecology often too readily associated with organic producer movements, and with ecological modernisation being associated with either industrial sectors and/or simplistic market-based assumptions of win-win environmental outcomes. I would contend, having taken the particular analytical journey displayed in this volume, that a focus upon rural development has much to benefit from the critical development of both bodies of theory. In addition, both bodies of theory, when examined critically can help to design a more effective and engaging rural development paradigm. One which progresses the intellectual and well as practical 'condition' of rural sustainability.

In the foregoing chapters of this last section of the book, moreover, I have been attempting to focus as much upon the obstacles now inherent in the rural development paradigm, as upon the real opportunities. In the chapter on short supply chains (Chapter 10) we see the evolutionary risks associated with alternative food supply chain developments; and in the case of organics in the UK (Chapter 11) we see that the overall contribution of the rise of organics and its relationships to sustainable rural development and agro-ecology are still contingent upon the reorganisation and regulation of supply chains more generally. There are, in short, obstacles and pitfalls in progressing *real* rural and sustainable development benefits, even where there is a growth of demand and supply of more sustainable food and other rural products and services.

These alternative rural development activities (as outlined in chapter 9) have to compete with the professional, economic, political and scientific logics explored in the earlier two sections of the book- the agro-industrial and the post-productivist logics. We have seen, for instance, that even when the post-productionist logic adopts a more environmental protectionist agenda (Chapter 8), the particular ways it does this both in compartmentalising the environmental problems and solutions and in restricting its appreciation of the role of agriculture, tends to continue to fragment the establishment of real and more viable alternatives from taking root in the countryside. In short we have to be critically aware of the rhetoric of sustainable development and ecological modernisation. And we have to apply an equivalent level of intellectual rigour to these discourses, as to those associated with the agro-industrial, post-productivist and post-modernist perspectives. This is a rich intellectual 'battleground.' To borrow Norman Long's phrasing: a 'battlefield of knowledge' (Long and Long 1992).

In this final chapter, I will explore a series of integrating concepts which begin to build a more rigorous and critical rural development perspective through the theoretical prism of what I term 'real' ecological modernisation.

Over recent years the sub-disciplines of rural sociology, development studies and the newly emerging social science of the environment have been attempting to address many of the same types of theoretical and empirical questions albeit from different starting points. The rural sociology of advanced economies has for a long time now drawn very productively from the field of development studies, even if both have remained quite distinct sub-disciplines with their own communities of interest and professional publication outlets. Similarly, the growth of work on the sociology of the environment, particularly in the European and North American context, has attempted to very much set its own broad agenda, but has done so in ways which have overlapped with the theoretical agenda of the more traditional rural sociology. The latter's increasing ecumenical concern for aspects of rural nature, as almost a 'last refuge' in the process of urban-based modernisation, has meant that it has been the originator of significant theoretical and conceptual impulses into the wider and growing debates which have encapsulated environmental social science. While the latter, it might be argued has taken on a more industrial and urban basis – addressing questions of ecological modernisation for instance – it has also needed to refer to key authors in the older, sister disciplines of rural and development sociology. Indeed, all three sub-areas witness the creative engagement of several key authors who have their origins in rural sociology and anthropology. None of these authors would just regard themselves as rural sociologists; rather they are contributing to wider sociological and geographical debates concerning nature, space and environmental policy (see, for instance, the contributions the Environmental Sociology Handbook, edited by Redclift and Woodgate 1998).

While many may suggest, in a tidy-minded fashion, that these developments represent something of a chaotic and disparate set of conditions which tend to diminish, rather than enhance, the true innovative character of rural sociology of advanced economies; I want, in this chapter, to demonstrate how I beg to differ. Rather, I want to begin to demonstrate how a wider theoretical and conceptual landscape can help to assist contemporary rural sociology in progressing its agenda. I note particularly here the recent assessment of rural sociology by Buttel (2001), where he makes the point that the past decades work on aspects of regulation and globalisation, for instance, has tended to be quite the reverse of the earlier developments of the 1970s and 80s (what was then termed the 'new rural sociology'). He argues for more theoretical innovation to be undertaken. While the past decade or so has demonstrated quite a flurry of rich empirically engaged work (on areas of globalisation for example), actual theoretical development has reached something of a hiatus; with various 'schools' tending to adopt a position along the actor-oriented political-economy axis. Moreover, as Buttel also alludes, this can lead to the rather random/chaotic choice of micro-empirical case studies; with the suggestion that this can begin to lose sight of 'the big picture'. A key question here then is how can a more theoretical agenda be established?

Ecological modernisation, and more sustainable rural development: towards new rural eco-realities.

However critical one might be about arguments associated with the rhetoric of sustainability, it is also the case that writers have pointed to the development, albeit fledgling in some cases, of a more ecologically modernising agenda. This is centrally associated with a European perspective on the development of clean industrial technologies, and the ways in which environmental coalitions and movements begin to affect reluctant state agencies. As I have suggested in the concluding section of this book, these ideas at least deserve some attention with regard to rural development. Indeed, the degree to which we might be entering a new phase of modernisation which would be much more ecological is becoming a central theme in rural and agro-food studies (see Goodman 1999, for instance); and the rural domain is becoming a central field for exploring the role and meaning of social nature debates. Also, as I argue above, many leading writers in the field are also contributing to the development of environmental social science, through publishing in their more recent bespoken journals and research collections.

Now taking Buttel's challenge seriously, and at the same time as the arrival of more effective debates about a new modernisation process, it is necessary in my view to address a central theoretical question. This in macro terms is: *to what extent are we seeing the arrival of a more ecologically modernising process operating in advanced societies, and as part of this, through rural development trends specifically*? Second, if we believe that this is a viable question, then what conceptual development and empirical realities does it suggest with respect to the rural sphere? Taking some empirical and theoretical avenues in the three resource areas of agricultural restructuring, food supply chains and forestry, I want to suggest that the answer to the first question is positive; and then demonstrate, through the definition of some conceptual parameters how we might address the challenge this creates for us in the second.[1] What I suggest we are witnessing is the development of an ecological modernisation process which is significantly different in its character from the early industrial modernisation process. As we saw in the first two sections of this book, the latter, which has by no means completely disappeared is, arguably, less in the ascendancy and is subject to internal and external crisis tendencies (such as food scares, legitimisation and ethical concerns), as well as deep structural tendencies which are seen as increasingly contradictory. Meanwhile social-ecological debates - albeit in a complex and diverse way - are gaining ground in different guises and through different types of social, political and economic practices. While these may confront the former structural changes, associated with globalisation of corporate capital, for instance, they are not, I will argue, simply marginal or subject to the marginalisation effects of corporate states, firms and their established social and political logics. Rather, they may suggest a new centrality for many of the features of rural life that the earlier industrially-based modernisation process tended to marginalise. For instance, aspects of agro-ecological development as part of rural development; the development of decentralised and more sustainable rural communities as a central part of settlement hierarchies; and more mobile and ICT-based sharing of experiences in rural and urban areas. Indeed, we might suggest that one central element of ecological modernisation is the very re-definition of the spatial and social balances between urban and rural living experiences, and about the re-alignment, more specifically, between nature, quality, region and locale producers and consumers for a more ecological rural resource base.

Some may see these notions as going too far. However, I think that we need to at least debate them in such a way as to re-build a more viable and robust rural development perspective which at least begins to suggest how such notions might be more effectively progressed. Indeed, just as with the industrial mode of modernisation, a new ecological paradigm also needs a viable, critical and normatively-engaging social science. This is particularly the case with EM given, as we shall see below, the spatially variable and context dependent ways in which it actually expresses itself. What seems clear is that there is a lack of coherence in EM, one which can be partially or contradictorily adopted and pursued by national, regional and local governments; and one which may need particular confluences of strategic and local interests and actors to operate in new and innovative ways.

Looking at the recent rural sociological, and particularly the environmental social science literature over recent years, one begins to see this tendency being reported. One further key question is how far and fast will it travel, and how do we best equip ourselves as scholars to develop a growing relevance in understanding and mediating its required and contested knowledges?

There are some large questions here. If we are to meet Buttel's challenge it would seem necessary that we have to start looking at these bigger pictures. The 'new rural sociology' and political economy that was established and developed throughout the 1980s and 1990's was built upon the development of an intense and very effective critique of the industrialisation modernisation project (particularly as it related to the agro-food complex). Now we should recognise that the challenge has begun to change; for it is no longer sufficient to critically examine the problems of such ageing 'regimes'. As I suggested in the introduction to the book, and then pursued at various points in it, we need to be reconstructing as well as de-constructing models and frameworks which suggest *how things could work in different and more socio-ecological ways over space and time*. We need to see better how actions and cases at one level can build up into significant projects of change more broadly.

In conclusion, here I will explore a series of key concepts and theoretical formulations which will hopefully begin to build not so much on the diffusion of effort that some might argue we have witnessed, but on some useful integrative themes which will, I believe, yield a more robust and theoretically engaging rural sociology with environmental sociology. This involves, I will argue, the need to re-problematise space and spatial relations (as socio-natural constructions) in our analyses, but to do so in ways which re-engage with the somewhat traditional concepts of *community, regulation, consumption, exclusion, justice, bureaucratisation, professionalisation and expertise, associationalism and responsibility*. By doing this I believe we can build upon the somewhat false, and dare I say now, somewhat futile and oversimplified dualisms between structures and actor strategies and networks, social constructivism and realism, globalisation and localism, economistically determined productionism and culturally confined consumptionism. I will argue that we need more engaging conceptual and theoretical formulations which help and guide us to not only make sense of the 'new ruralities' which confront us, but also allow us to play a more *reconstructivist* as well as *critical interpretive* role in both rural and environmental social science.

Key conceptual starting points: empowering social ecologies

Most recently, I have been engaged in three rather long term projects, which while linked have concerned themselves with quite distinct areas of activity and arenas of natural resource.[1] These are associated with agro-food developments, farming and rural development, and thirdly, and more recently with the forestry sector. In conducting this work with a variety of colleagues I have been made increasingly aware how aspects of what we might term, ecological modernisation are contestably taking hold; albeit interfacing and competing with more established and conventional modernising models. New alternatives, for instance, in food supply chain development – socio-natural networks – are clearly establishing themselves which I would argue are new (although I recognise that a conceptual and empirical challenge here is very much to persuade some of our close colleagues that this is little more than 'old wine in new bottles'). Similarly in the farming field, detailed investigations in England and Wales, as well as in other European regions, exhibit a variable shift toward what we have termed the 'broadening, deepening and regrounding of conventional agricultural practices'. In the forestry sector too, with the slow adoption of common property resource concepts in some European regions, and even in the UK now some recourse to sustainability and social inclusion discourses, we see the traditional, narrow definition of the forestry as productivist rural space beginning to be challenged (Mather 2001; Bishop *et al.* 2002).

Emerging out of these three theoretical and empirical explorations we see, moreover, the need to conceptually progress a rural and environmental sociology which can accommodate and facilitate ecological modernisation. What are then some of these conceptual and theoretical coordinates for a more socio-natural rural sociology? I outline here six such theoretical coordinates making reference to the three rural resource areas of food, farming and forestry.

Environmental and territorial justice

One key feature of current patterns of rural development concern what seems to be greater amounts of economic and geographical uneven development. This is partly exacerbated by the gradual shift, on the part of many nation states, away from explicitly fostering more convergent, balanced, regional economic development. A key feature of extant rural and environmental arenas comes then in needing to address aspects of both territorial and environmental justice. The adoption of sustainable development goals cannot really be progressed unless it begins to deal with spatial uneven development, and particularly how local and regional people's conditions are incorporated into the analysis of how those broader and long term conditions could be improved. A central question to consider, is the extent to which scholars engage with the debates and assumptions of territorial and environmental justice. That is, having identified injustices in the uneven risk, distribution and quality of resources in space, how far should we go in formulating ways of reducing or compensating for this unevenness (see SDRN 2002)? For instance, it is quite clear from our work on forestry and community in South Wales that any attempts by the Forestry Commission to encourage more participation and inclusion in the forested areas is constrained by the quite long history of growing social and spatially based deprivation of the nearby communities, for which the withdrawal of other state services and support structures has played a full part (see Marsden *et al.* 2001b). Similarly, the growing social science research in the areas of food scarcity, and the identification of 'food deserts' – often themselves hidden enclaves in larger, prosperous cities and regions – demonstrates, on the one hand, the acute unevenness of rights to food consumption that the now corporate retailer-led supply chains have reproduced (see Carley *et al.* 2001). The question is, however, not simply the issue of the simple empirical identification of these disparities. It also concerns how an engaging social science of rural development and food formulates alternatives which reduce and combat these. In the rural development field, while many of the grant-aided funding schemes emanating out of the EU still have a strong territorial justice philosophy behind them (e.g. Structures funding, LEADER etc), the emphasis of a social capital and capacity logic governing local and regional allocations systems inevitably means that many rural groups and localities are failing to develop the critical mass to establish and capture such funding benefits.

Hence, we can see here that progressing environmental and territorial justice questions are almost unavoidable in addressing these new developmental trends. It these cases we need to do so explicitly. For instance what are the consequences of the further concentration of resources and economic development? What sort of trade-offs are possible or legitimate? How can injustices be compensated, monetarily and non-monetarily? How can a more multi-level governance system in the rural development sphere (i.e. combinations of EU, national, regional and local policy platforms) progress a renewed set of regional and local convergence policies? Moreover, what are the social and political effects of persistent injustices in and through different rural spaces?

Community and Association

Much of the social science of the environment literature (see Irwin 2001) and rural social nature debates, for instance, have tended to so far underplay one of the traditional strengths of rural sociology: community. It is somewhat surprising then that notions of community and association have so far failed to intervene as important working and active mechanisms between the 'social' and the 'natural'. Two examples illustrate this from our work on farming and rural development, and from the forestry and community work.

Alternative supply chains and rural development activities are, as we saw in Chapter 10, it built upon new sets of associations and associational capacities of actors to engage in ways which shape both the social and the natural. Associational interfaces (often as emerging and contingent networks and chains) are both informal but highly significant in establishing trust, common understandings, working patterns, and forms of co-operation and co-optation between different actors in the supply chain. These differ from institutional interfaces which include state regulations and the support and services offered by rural economic development agencies (see Long 2001). And, they are often crucial in generating and facilitating supply chain interfaces at a regional level – a key new spatial platform for the development of agro-food alternatives to take hold (Chapter 9). However, such interfaces are vulnerable to internal and externally generated disruptions. There is no inevitability that strong and mutually reinforcing associational interfaces will be reproduced over time. Furthermore, where they do not exist, or have broken down, it may take many years to re-build relationships and working trust relationships to a point where regional or local actors – or actors operating at a distance across a supply chain – can create the conditions necessary to effectively and efficiently meet and relate to consumer demands. In many cases, in what I have termed 'short supply chains' these new associational competences are formed across the traditionally recognised producer/consumer arena. Indeed, re-connecting associational capacity across this former divide is part of a new set of innovations that mark out the new modernisation projects from the old. Our evidence from European case studies suggests that sustaining rural development through the evolution of reconfigured supply chains must be based upon new combinations of both institutional support and associational development. Furthermore, these relationships must be able to adapt to internal and external shocks and pressures over time and space. Here, and again distinctively, there is no one model. A key question then is the degree to which these features will need to become more widespread if real aggregated rural development impacts are to be achieved, and how, if they are not apparent, could they be generated?

In the local context of residential forest communities in South Wales, we also see how deep-seated local interpretations and impressions of community, and particularly of the marginalisation effects of the State as part of community life, have a critical bearing upon the constructions and shapes of socio-natural relations and practices in that community. In short, the social character and construction of the community – in this case a socially rich place, but marginalised and peripheralised by the State – provides an important social prism through which nature is perceived and used (see Bishop *et al.* 2002). Hence socio-natural perceptions and practices emerge as an expression of the social construction of the community; and in this case, the particular incision of the state in this construction. In this sense what we can infer from this is that particular types or sets of socio-natural relations and practices cannot be simply 'read-off' from the prevailing sets of *realist* social and economic regulatory conditions. Neither, however, as social constructionists may propose, can they just be rendered as varyingly 'co-constructed' within environmentally related practices and social contexts' (Irwin 2001). The residents in the community are not just free agents in their constructions of natural relations. Neither is the forest a neutral natural actant operating in a completely natural setting. Both nature (in this case the forest) and the people are actively engaging in broader socio-political conditions which, in turn help to shape the particular types of socio-natural relations in this type of local context. Such a postulate suggests that there is an inherent partiality about the two earlier approaches. What needs to be added to both conceptions is an empirically embedded understanding of how the natural becomes an active ingredient in the continual process of community construction.

Exclusion and empowerment

While aspects of rural social exclusion have become a vibrant topic of concern, especially amongst rural geographers (see Cloke *et al.* 2000; Shucksmith 2000); the emphasis of this work has thus far been largely upon the identification of exclusionary practices in different social, economic, institutional and domestic arena. One further and important aspect – as agro-ecological frameworks suggest – concerns the ways in which communities, different actors and agencies can become more empowered; and to do so in contexts which are ecologically as well as socially sustainable. In our forestry communities we witness a long historical process of disempowerment brought about by both the productivist priorities of the mining and forestry sectors, and in addition, the gradual marginalisation of much of the community by labour markets and the state authorities (see Chapter 6). Here disempowerment is multi-faceted, and it is the combination of these marginalisation histories which provide the backdrop for any change of Forestry Commission policy in terms of enhancing community and sustainable development. The exclusion-empowerment equations operate differently between and through communities and they provide a key mechanism in establishing the social landscape upon which new, potentially more ecological forms of modernisation could take hold. Indeed, they provide a major way in which ecological modernising principles can in themselves become marginalised and obscured in the face of what are seen to be more short-term employment imperatives.

A further, if separate element of exclusion and empowerment comes in terms of the market and consumer exclusions operating in the conventional food supply system. Here we see two key processes at work which deserve much more attention in progressing a more ecologically modernising agenda. These concern the exclusionary tendencies based both upon price, quality and location encouraged by the corporate retailers in the operation of their food chains (see Competition Commission 2000). This is generating a major feature of the 'new food politics' with consumer agencies adopting exclusionary arguments as well as food quality arguments in their political articulations with government authorities. The operation of what is called in the trade as 'price flexing' – whereby the corporate retailers purposely vary their grocery prices in the light of local competitive conditions rather than being related to actual costs- has become widespread in the UK. Even the conservatively written recent Competition Report of the national government concluded that where the big corporate retailers were practising this, it tended to operate against the public interest. Customers tend to pay more at stores that do not face particular competitors than they would do if those competitors were present in the area. In many of the marginalised working class areas of urban and rural UK these monopoly retailing arrangements are now common place. Hence we see here that there is a real need to link the particular corporate and institutional practices with local empowerment and exclusionary conditions – in space. In many of the debates so far – both academic and public policy – these linkages in space have been at best blurred. Indeed, the spacing of exclusionary practices and their links to corporate strategies has been neglected. It is not so much about the identification of 'food deserts' as understanding the interfaces of corporate strategy and local consumption in their varied spatial contexts. As this example shows, what we witness in many areas are competition practices going in reverse. With major retailers being able to create almost 'monopoly consumption spaces' in which they can raise prices, beyond their relationships with 'true' costs. Keeping other retail competitors *out* as well as harnessing loyal and recurrent consumers *in* to their stores becomes a central mechanism *in* the territoriality of exclusive food consumption in the British context.

A third arena of exclusion/empowerment concerns those at the other end of the food supply chain-the primary producers. Despite government policy rhetoric about the need to abide by the laws of fair trading and European competition policy, many farmers and many processors find themselves excluded from the often more lucrative retailer – led food markets. For those that do gain entry, the degree of informal control over their operations severely constrains their 'room for manoeuvre'. In many ways these are the newly created forces of subsumption operating in the retailer-led chains; whereby pricing and conditions can be placed upon producers and processors in ways which are seen to be 'consumer-led', but really emerge from the day-to day management of these supply chains by the category managers of the retailers. With exclusion or relative empowerment in this regard, therefore, we see also the creative development of social and economic dependencies operating between different sets of actors in the supply chain. Local rural and agricultural spaces are, therefore, no longer controlled at a distance by the corporate input suppliers (i.e. associated with concepts of appropriation, substitution and subsumption of production). Now it is more variably conditioned by the down-stream retailers who engage groups of consumers and governments in their powerful networks. Hence, primary production becomes something of a backwater and 'dirty' activity which needs to be 'cleaned up' by the state and corporate retailers on behalf of the public.

We see here, then, through these examples how exclusion and empowerment strands are a critical integrating mechanism for bringing together the social, economic and institutional construction and allocation of power, both in and through rural (as well as urban) spaces. This is cross-cutting, both vertically through supply chains and laterally through community/ institutional interfaces.

Consumption and production: socially and culturally reconstructing the 'quality' commodity

While considerably more emphasis has been placed upon understanding the relationships between production and consumption; and particularly the ways in which the very fabric of food fuses both natural and social hybridities, there needs to be more attention given to the ways in which the hybrid social and natural relations surrounding food are governed, empowered and used. While accepting that there has been something (although I think somewhat overdrawn) of a conceptual distinction made between the more political economy-inspired analysis of production and the more culturally-inspired consumption studies, there is a clear need to explore how new alliances and relationships can be brought together in ways which progress the understanding of alternative food networks or chains. Also, we need to know much more about how these both use and travel through space. Our recent European evidence suggests high levels of spatial variability between different rural regions in the setting up and sustenance of alternative supply chains. These variabilities are particularly affected by the type and degree of institutional support, via, not just the national state, but regional knowledge and skill-based agencies; and by the highly context dependent types of associational involvement from a variety of different actors. It is clear that the innovation patterns, the skills bases and the degree to which such initiatives can utilise and exploit the 'territorial worth' of their localities and regions are all critical factors. These factors show that aspects of space, quality convention, nature, and agricultural and food socio-technical practices (e.g. the particular ways of cutting, curing, salting and carving of meats, for instance) come together in rural spaces. Moreover, many of these initiatives also reconstruct important new co-locational alliances with specific groups of consumers. Many consumers, for their part are increasingly searching for 'something else' other than the standardised or indeed mass-produced specialised types of food product. The consumption experience, often involving direct contact with the producer or the seller becomes an important ritualistic consumption practice.

However, what marks these alternative food chains out from the conventional system, is by no means their face-to-face nature necessarily. In some of the more mature quality supply chains we see the development of spatially extended networks, which are selling brands, labels and seriously commodifying their culinary repertoires (e.g. Parmigiano Reggiano Cheese). They are still categorically alternative, however, in that that have and do re-align nature, space, socio-technical practices, and quality conventions in ways which make it impossible to replicate these outside that network. These then are the new ecologically-deepened supply chains, and to describe and understand them requires much more than a recourse to such generalised notions of mutual or reciprocal hybridities or metabolics. Rather, it requires a concerted effort to understand the *theory and practices of the actors* – both producer, consumers and exchange actors – who mobilise and animate/demarcate such networks (see Verschoor 1998, for instance). These new food networks are now a common occurrence and in many regions are by no means marginal to the conventional system. In Northern Italy, we see a concentration of quality food networks, while in Germany and the Netherlands, new alliances are more associated with agri-tourism ventures and nature and landscape management. Food as a culturally hybrid entity, which becomes unevenly embedded into the fabric of new rural development practices; and new synergies become developed between food, agricultural practice, consumption practices and associational and institutional arrangements.

Understanding the wealth and worth of these series of micro-social practices, and still more, assessing what in more generalised terms they 'add up to' is a major empirical and theoretical task to which Buttel (2001) alludes, and for which scholars have only recently embarked. It is clear, however, that the volatilities in the consumption/production relations associated with the conventional systems, will have pervasive effects and interactions upon the competitive spaces that these alternative food supply chains operate in. This is highly contingent, not just upon some vague notions of consumer culture or sovereignty. It is also centrally conditioned by the contested and institutional ways in which the 'consumer interest' is continually constructed – and by whom. Such analytical progress will not simply rely upon a realisation that nature and society are hybridised in food. The issue is how, where and by which powerful/ non-powerful actors? Food as a realm of governance and social and political regulation is not contradicted by more socially constructivist processes. These need to be put together; not least because institutional actors and agencies are part of that constructed and contested process. Somehow, we need to understand how alternative food networks not only get formed, but then get demarcated and maintained as a socio-competitive dynamic. They are not static events.

Corporate responsibility and accountability

The past two decades have witnessed a significant growth in research work on the agro-industrial globalisation and regulation of agribusiness and the workings of multi-nationals (see Friedland 2001 for review of the commodity systems approach). This has been a major development of the discipline of rural sociology, and it has emerged as a clear critique of industrial modernisation in the agro-food sector. In postulating aspects of ecological modernisation, however, we also now need to consider and critically assess how business relationships (both with each other, with consumers, and with local communities) are changing as a result of a number of significant forces. These include: changes in the regulatory frameworks; the globalisation of supply chains and the increasing mobility of capital and knowledge; the application of ICT to radically restructure certain markets, industries and places; changing patterns of consumption and falling consumer trust in existing forms of corporate and government communications; the emergence of civil society groups and global protest movements and a realisation that companies may be more vulnerable to pressure for change than entrenched or undemocratic governments; the fragmentation of local communities and constructions of community; the pressure from the investment community in the form of the growing ethical investment movement; and the pressure from insurers who see the reduction of environmental damage and social conflict as an important way for companies to reduce levels of risk.

These reconstituting forces are, in turn, reshaping the relationships between firms, governments and 'communities of interest'. Amongst the actors there are key constituencies who will help to shape the dynamic and developing relationships both within their actor-spaces and between themselves and other actors. From the point of view of business it is now more essential (albeit to varying spatial and sectoral degrees) that these changing relationships are constructed and constantly reconstructed in ways that help to ensure its longer-term survival. These can be depicted in the Figure 12.1. However, neither the effects on businesses of the processes of change that I identify here will not be even, and nor will be the impacts on the 'communities of interest'. *What is necessary in research terms then is to theoretically and empirically understand the richness, complexity and contestability of business responses to, and the management of, its wider environment.*

Figure 12.1 Changing relationships between key actors

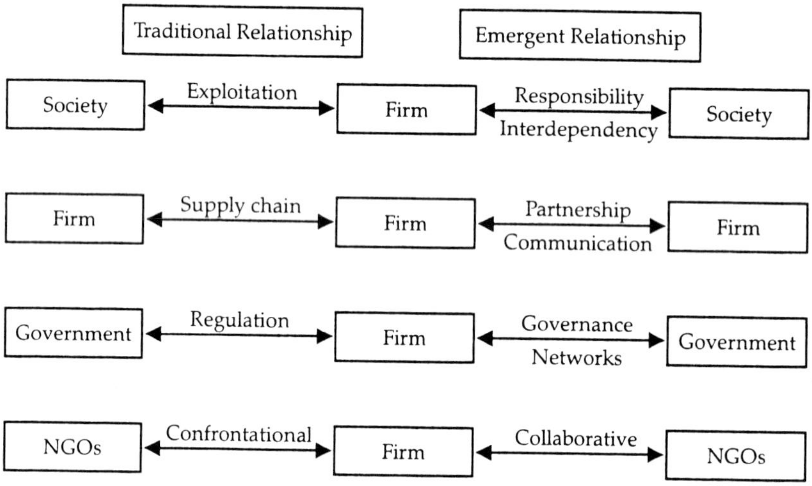

In short, we need new constructed models of business behaviour and relationships which are embedded in the social, environmental and political contexts, which they are also attempting to shape. So far environmental and rural sociologists have tended to avoid these ventures; leaving it to the environmental economists and ethical marketing specialists to engage in the tracing of these new patterns. For instance, such discipline-specific research is pursuing the following lines:

- the inclusion of sustainability and social responsiveness within corporate accounting and reporting practices, and its relationships to the growing ethical investment industry is typically being researched within the accounting and finance disciplines (e.g. Gray *et al.* 1995).
- changing consumer demand for ethically and environmentally acceptable products and the response of companies to demands for such products, new information and new marketing approaches is typically being researched by the marketing and economics disciplines (e.g. Menon and Menon 1997).
- the evolution of environmental and social regulation and its impact on firms is typically researched within the legal and public policy discipline (e.g. Garrod 2000)

- the implications of sustainability in terms of industry structure and infrastructure is typically a focus of research in the planning, geography and public policy disciplines – and, the influence of global capital flows and governance standards is primarily researched within the law, finance and international relations disciplines (see Brack 1995).

Yet, in all these areas environmental social science, and more specifically rural social science, has a potentially important role to play, particularly in understanding how these changing business, government, community relationships are spaced and re-spaced. In the food sector, for instance, the contested competitive spaces between conventional and alternative supply chains is occurring in a broader context of a reduced reliance simply upon legal compliance as a justification and legitimisation of actions. In this sense de *jure* regulation is not enough. Also there is a broadening amongst many firms of the 'stakeholder concept' whereby companies are increasingly having to respond not just to customers and investors, but also those with a physical stake in their actions, including the communities within which they operate and NGO's representing a myriad of interests (Polonsky 1995). We see this empirically, for instance in the redefinition of UK forestry commission policy and the new sets of relationships developing with government agencies. Moreover, consumer groups and governments have traditionally championed consumer rights in terms of choice, value and safety of products. Yet concerns over the sustainability and social responsiveness of businesses has led to an increased interest among stakeholders about the methods and consequences of production (Peattie 1999). This is most easily observable within the food industry, where concern is centred on issues relating to production practices – such as the use of chemicals, the use of GM and the constituents of animal feedstuffs, pollution and the destruction of habitats.

These trends suggest, at the very least a set of key research questions:

- how are changing patterns of demand leading to the development of more sustainable and socially responsible strategies, products and technologies?
- what types of new market and institutional structures (e.g. supply loops and networks) will develop in response to product take-back, as businesses are required by legislation and consumer pressure to become even more accountable for the life-cycle of their products?
- what opportunities will arise for the restructuring of industries around new business models and types of production systems and technologies?
- how will business to business and business to consumer relationships change if a service based economy (i.e. through leasing, take-back and

service arrangements) begins to replace the traditional exchange of physical goods?

To begin to answer some of these questions will require the development of new social and regulatory 'designs' around which we can begin to explore the richness and complexity of business/ environment relationships. Such an endeavour will need to take us far beyond the much more established and somewhat retrospective 'tracking' of the agro-industrial multi-national commodity flows. Moreover, we should put aside the rather binary question of whether firms will or will not ecologically modernise. Rather, we should explore both *what is* and *what might be.*

Regulation and bureaucratic professionalisation

As I argued in Chapter 8, it is important to recognise that the story of environmentalisation over the past two decades in the rural field has been one in which a particular form of Webarian bureaucracy has been increasingly prevalent. We need to recognise, both why this has been the case, and conversely, why it is not necessarily the only model in town. In the context of the agricultural and food sector in Europe (post BSE) termed this more specifically the *hygienic mode of regulation.* As the industrial mode of food supply has become even more crisis-ridden, the state has attempted to largely 'correct' this by setting up highly professionalised and bureaucratised forms of environmental safeguards and instruments. This has also been conducted in ' the interests of the consumer', as a way of governments seeking to protect their interests at the same time as allowing corporate industry the ability to exploit new markets. For instance, one recent expression of this is the advocation by some corporate supermarkets and government officials that eventually *all* farmers will have to old some form of quality licence to operate. The growth of a profound regulatory burden, as a response to the crisis in the industrial mode of agro-food, tends therefore, only to strengthen the economic and political power of established agro-industrial interests (including the large retailers and manufacturers). Both private and public forms of regulation are used to 'clean-up' the industrial system in the ostensible 'public interest'. Such *schematisation* holds the added consequence of further constraining the real potential of integrated agricultural development as well as providing new regulatory barriers to market entry for many smaller producers and processors. For consumers, it allows the disconnections and distanciations between production and consumption to conveniently continue: with an encouragement that 'safety' comes before (or gets conflated with) sustainability.

These processes have tended to break, or at least 'fracture' the environmental and rural development question into specific boxes, projects and schemes; making it more difficult to make holistic connections (as associated with agro-ecology for example ; see Guzman and Woodgate 1999). As a result it has been difficult for many actors to construct viable and integrated alternatives, or to harness the necessary spatial, natural, regional and knowledge-based resources necessary to progress *real* rural development options.

The broader field of rural development, as well as that of agro-food, has also become increasingly populated with project managers, consultants, exchange agents etc. such that a profession has been established to which some are excluded. Moreover, current priorities of national governments as well as emerged EU food quality policy concern the reconciling of the demands and risks of the 'careful consumer'. In the agro-food sector this is largely done by assembling a bureaucratic-hygienic apparatus-itself something of a new compromise between governments and capital-in order to stave off further and potentially deeper consumer-led legitimisation crisis in the old industrial system as a whole.

There is a need, therefore, for rural social scientists to contest these regulatory modes and to apply other, for instance, agro-ecological and food supply-chain ecological models to rural realities. This will require us to explore the contradictions and practices embedded in the different modes of environmental policy discourse, and for us to challenge the specificity of environmental expertise and professionalisation. How can a more holistic food ecology be created? And, what novel forms of regulation would this require? Is the bureaucratic/hygienic mode sustainable over time and space? And what barriers does it place upon achieving real ecological modernisation? For instance, as I argued in Chapter 1, while the current agro-environmental policy and rural development discourse continues apace in the European Commission, at the very same time we see, for instance in Finland, the rapid structural concentration of farms occurring as a result of continued policies of CAP. The surface waves of ecological modernisation discourses often seem to obscure the deeper countervailing currents of structural changes in the agro-food sector.

Conclusions: looking through the environmental maze and broadening politial economy

Past analyses in rural, and to a large extent in environmental sociology more generally, have to a great extent been built upon sophisticated and critical interpretations of the industrial modernisation project. In rural development terms, for example, this has focused upon challenging the economistic notions of scale, critical mass, centralisation, globalisation and marginalisation that this model has clearly engendered. Ecological modernisation, in all of its shades, nuances and theoretical flaws, at least brings forth a new question. That is, how could/should the relationships between civil society, the state and the market be re-arranged in ways which would usher in different types of more sustainable development which would incorporate ecological worth? A 'new age' of ecology, however, has begun in a slow, fledgling and uneven way. I have argued here that it does have some real intellectual purchase and potential, that part of this allows us to overcome some of the theoretical rigidities of the past, and that part of its value is that it brings spatiality into a central arena in rural political economy. In addition, it opens the door to more theoretical pluralism, debate and progress (as I have tried to explore in the three sections of this book); moving beyond social constructivist versus realist debates, structure and action and macro and micro. Here, in conclusion, I have attempted to set out here some of the cardinal conceptual reference points which have emerged as I have conducted three different research projects in the food, farming and forestry sectors. These are then partly empirically grounded as well as theoretically constructed arguments and, as such they deserve much more expansion, debate and consideration than is feasible here.

However, one of the fruitful opportunities such an exercise begins to explore, is the possibilities of re-assembling conceptual frameworks in ways which integrate nature and space within the broader political economy of national and international ruralities. As I have argued above, many of the current environmental discourses and regulatory frameworks tend to fracture such integrative endeavours in such ways as to create false divisions of professional labour between different communities of interest (including academics, policy officials environmental NGO's, consultants, and a whole range of 'project managers' who are now engaged in competitively progressing rural development initiatives). This makes integration in policy development and implementation all that more difficult, and it leads to the driving out of a radical consideration of alternatives by the urgency of meeting bureaucratically applied performance measures and indicators. Hence the critical development of environmental and rural social science is now a more central task in creatively engaging with what might be, as well as what is. The conceptual starting points I have outlined here are middle level concepts which can start to address this task. We should remember that the agro-industrial modernisation project has developed an elaborative and justificatory social science with which to legitimise itself; yet an equivalent social science for ecological modernisation is still very much in its infancy and liable to marginalisation. In undertaking these explorations it will be necessary, as most of the conceptual parameters suggested here demonstrate, to embed our conceptions of 'producers' and 'consumers' much more into the fabric of the contested worlds of industrial and ecological modernisation currently being applied in the advanced economies.

Note

The arguments developed here draw upon three research projects being conducted by the author and several colleagues (including, Jo Banks, Paul Milbourne, Jon Murdoch, Kevin Bishop). The first was entitled *Innovation and Quality in the Food chain: strengthening the regional dimension'* (funded by the Economic and Social research Council). The second is entitled *'The Socio-economic impact of rural development policies: realities and potentials'* funded by the EU; and the third related to an ethnographic study, *'Forestry, community and land in the South Wales Valleys'* funded by the UK Forestry Commission. The early results ideas emanating from these projects are contained in recent articles (see Murdoch *et al.* 2000; Marsden *et al.* 2001b; Bishop *et al.* 2002).

References

ABRAM S., MURDOCH J., MARSDEN T.K. (1996) The social construction of 'Middle England': the politics of participation in forward planning. *Journal of Rural Studies* 12, pp. 353–364.

ADAM B. (1999) Industrial food for thought: timescapes of Risk. *Environmental Values* 8, 219–238.

ADAS (AGRICULTURAL DEVELOPMENT AND ADVISORY SERVICE) (1996) *Tir Cymen Socio-economic Assessment Final Report*. Research Report Prepared by ADAS for the Countryside Council for Wales (ADAS, Aberystwyth)

AEBC (AGRICULTURE AND ENVIRONMENT BIOTECHNOLOGY COMMISSION) (2002) *Looking Ahead. An AEBC Horizon Scan* (AEBC, London).

AGRA-EUROPE (1994) Domination of food retailing by supermarkets: a world wide trend *Food policy international* 3(2), pp. 2–4.

AGRA-EUROPE (1998). Labour and land costs handicap EU dairy farms, August 1998

ALTIERI A. (1987) *Agro-ecology. The Scientific Basis of Alternative Agriculture* (Westview Press, Boulder, CO).

AMIN A., THRIFT N. (1995) Institutional issues for the European regions: from markets and plans to socio-economics and powers of association *Economy and Society* 24, 41–66.

ANDERSON M., MASSA I. (2000) Ecological modernization: origins, dilemmas and future directions *Journal of Environmental Policy and Planning* 2, pp. 337–345.

ANDRADE M.C. DE (1982) *Tradição e Mudanga* (Zahar Editores, Rio de Janeiro).

APPADURAI A. (1986) *The Social Life of Things. Commodities in Cultural Perspective* (Cambridge University Press, Cambridge).

ARAÚJO T.B. DE (1995) Nordeste, nordestes: que Nordeste? In: AFFONSO R. DE B. A., SILVA P.L.B. (eds) *Desigualdades regionais e desenvolvimento. Federalismo no Brasil* (Fundap, São Paulo) pp. 125–156.

ARAÚJO T.B. DE (1997) Heranga de Diferenciação e Futuro de Fragmentação *Revista do Instituto de Avangados da USP Estudos* 11(29), pp. 7–35.

ARCE A., MARSDEN T.K. (1993) The social construction of international food: a research agenda *Economic Geography* 69, pp. 293–312.

ARKELTON TRUST (1993) *Farm Household Adjustment in Western Europe 1987-1991*. Final report on the research programme on farm structures and pluriactivity (European Commission, Brussels).

ARTON WILSON COMMITTEE (1956) *Report of the Committee to Review the Provincial and Local Organisation and Procedures of the MAFF* (London, HMSO).

ASA (Advertising Standards Authority). 2000. *Adjudication Soil Association July 2000* (http://www.asa.org.uk).

ASHBY A., EVANS H. (1944) *The Agriculture of Wales and Monmouthshire* (University of Wales Press, Cardiff).

BANKS J. (2000) *Organic Food Supply Chains in Wales. Impact and Potential* (Department of City and Regional Planning, Cardiff University, Cardiff).

BANKS J. (2001) *Farm Retailing in the UK*. Papers in Environmental Planning Research (Department of City and Regional Planning, Cardiff University, Cardiff).

BANKS J. (forthcoming). *Ecological Dairy Farming in West Wales*. Working Paper (Wageningen University, Wageningen).

BANKS J., BRISTOW G. (1999) Developing quality in agro-food supply chains: a Welsh perspective *International Planning Studies* 4, pp. 317–332.

BANKS J., MARSDEN T.K. (1997) Reregulating the UK dairy industry: the changing nature of competitive space *Socilogia Ruralis* 37, pp. 382–404.

BANKS J., MARSDEN T.K. (2000) Intregrating agri-environmental policy, farming systems and rural development: Tir Cymen in Wales. *Sociologia Ruralis* 40, pp. 466–480.

BANKS J., FOSTER C., LAMPKIN N., PADEL S., PARROTT N. (2000) *Organic Production, Consumption and Consumer Demand in North Wales* (Department of City and Regional Planning, Cardiff University, Cardiff).

BECK U. (1992) *Risk Society. Towards a New Modernity* (Sage Publications, London).

BELL R. (1999) British food retailing: a comparative analysis *European Regional Review* pp. 30–35.

BENDINI M., BONACCORSI N. (1998) *Con Las Puras Manos* (Editora La Colmena, Buenos Aires).

BISHOP K., KITCHEN L., MARSDEN T.K., MILBOURNE P. (2002) *Forestry, Community and Land in the South Wales Valleys*. Final Report to the Forestry Commission (Department of City and Regional Planning, Cardiff University, Cardiff).

BONANNO A., BUSCH L., FRIEDLAND W.H., GOUVEIA L., MINZONE E. (eds) (1994) *From Columbus to ConAgra. The Globalization of Agriculture and Food* (University of Kansas Press, Lawrence KA).

BOYLE P., HALFACREE K. (eds) (1998) *Migration into Rural Areas. Theories and Issues* (John Wiley & Sons, Chichester).

BRACK D. (1995) Balancing trade and the environment *International Affairs* 71, 497–514.

BRASIL. CODEVASF (1996) *Projetos de Irrigação no Vale do Sio Francisco*. Brasilia.

BRISTOW G. (1997), *Making the CAP Fit. An Integrated Development Strategy for Rural Wales* (Institute of Welsh Affairs, Cardiff)

BUCK D., GETZ C., GUTHMAN J. (1997) From farm to table: the organic vegetable commodity chain of northern California. *Sociologia Ruralis* 37, pp. 3–19.

BUTTEL F. (1997) Some observations on agro-food change and the future of agricultural sustainability movements. In GOODMAN D., WATTS M. (eds) *Globalising Food. Agrarian Questions and Global Restructuring* (London: Routledge) pp. 344–365.

BUTTEL F.H. (2001) Some reflections on late twentieth century agrarian political economy *Sociologia Ruralis* 41, pp. 165–182.

CABINET OFFICE (1999) *Rural Economies* (The Stationery Office, London).

CALLON M. (1986) Some elements of a sociology of translation: domestication of the scallops and the fishermen of Saint Brieuc Bay. In LAW J. (ed.) *Power, Action and Belief. A New Sociology of Knowledge?* (Routledge & Kegan Paul, London) pp. 196–233.

CARLEY M., KIRK K., MCINTOSH S. (2001) *Retailing, sustainability and neighbourhood regeneration* (Joseph Rountree Foundation, York).

CAVALCANTI J.S.B. (1996a) Globalizaçao, urbanizaçao e reprodu95o da forga de trabalho: politicas energ6ticas e irrigagio no Vale do Sio Francisco. In MAGALHAES S., CASTRO E.R. (eds) *Energia na Amazónia*, vol. 1 (Museu Paraense Emilio Goeldi, Belem) pp. 421–499.

CAVALCANTI J.S.B. (1996b) Globalização, novas regiões de produção agrícola e desigualdades sociais *Caderno do CRH* 24/25, pp. 109–121.

CAVALCANTI J.S.B., RAMOS J.V.R., SILVA A.C. DA. (1998) El trabajo femenino en la agricultura de exportación. Las trabajadoras en la producción de uva Brasil. In BENDINI M., BONACCORSI N. (eds) *Con Las Puras Manos* (Editora La Colmena, Buenos Aires) pp. 77–94.

CCW (COUNTRYSIDE COUNCIL FOR WALES) (1997) *Tir Cymen. The First Five Years* (CCW, Bangor).

CHRISTOPHERSON S. (1993) Market rules and territorial outcomes: the case of the United States *International Journal of Urban and Regional Research* 17, pp. 174–188.

CLARK G. (1992) 'Real' regulation: the administrative state *Environment and Planning A* 24, pp. 615–627.

CLOKE P., THRIFT N. (1990) Class and change in rural Britain. In: MARSDEN T.K., LOWE P., WHATMORE S. (eds) *Rural Restructuring. Global Processes and Local Responses* (David Fulton Publishers, London) pp. 165–181.

CLOKE P., MILBOURNE P., WIDDOWFIELD R.C. (2000) The hidden and emerging spaces of rural homelessness *Environment and Planning A* 32, pp. 77–90.

COMPETITION COMMISSION (2000) *Supermarkets. A Report on the Supply of Groceries from Multiple Store in the United Kingdom* (3 vols) (HMSO, London).

COOK I., CRANG P. (1998) The world on a plate: culinary culture, displacement and geographical knowledges *Journal of Material Culture* 1, pp. 131–153.

COOMBES B., CAMPBELL H. (1998) Dependent reproduction of alternative modes of agriculture: organic farming in New Zealand *Sociologia Ruralis* 38, pp. 127–145.

COUNTRYSIDE COMMISSION (1995) *Evidence to the House of Commons Environment Committee. The Environmental Impact of Leisure Activities* (Countryside Commission, Cheltenham).

CUMBRIA COUNTY COUNCIL (1997) *Rural Restructuring and Socio-economic Development in Cumbria* (Department of the Environment).

DATAMONITOR (1999) *Natural and Organic Food and Drinks 1999* (Datamonitor Europe, London).

DAVIES J. (1974) The end of the great estates and the rise of freehold farming in Wales *Welsh History Review* 7, pp. 186–211.

DAY G, REES G, MURDOCH J. (1989) Social change, rural localities and the state: the restructuring in rural Wales *Journal of Rural Studies* 5, pp. 227–244.

DRUMMOND I., MARSDEN T.K. (1995) A case study of unsustainability: the Barbados sugar industry *Geography* 80, pp. 342–354.

DRUMMOND I., MARSDEN T.K. (1999) *The Condition of Sustainability.* (Routledge, London).

DRUMMOND I., CAMPBELL H., LAWRENCE G., SYMES D. (2000) Contingent and structural crisis in British agriculture *Sociologia Ruralis* 40, pp. 111–128.

DTZ PIEDA CONSULTING (1999) *UK Speciality Food and Drink Sector, 1999* (DTZ Pieda Consulting, Edinburgh).

ENTICOTT G. (2000) *Heterogeneous Ruralities. The Place of Nature and Community in the Differentiated Countryside.* Unpublished PhD Thesis (Cardiff University, Cardiff).

ESRC (ECONOMIC AND SOCIAL RESEARCH COUNCIL) (1999) *The Politics of GM Food. Risk, Science and Public Trust* (ESRC, Swindon).

FANFANI R. (1994) Agricultural change and agro-food districts in Italy. In SYMES D., JANSEN A.J. (eds) *Agricultural Restructuring and Rural Change in Europe* (Wageningen Agricultural University, Wageningen).

FEIERMAN S. (1990) *Peasant Intellectuals. Anthropology and History in Tanzania* (University of Wisconsin Press, Madison, WI).

FESTING H. (1998) Farmers markets in the UK *Food Policy* 28.

FITZSIMMONS M. (1989) The matter of nature *Antipode* 21, pp. 106–120.

FLYNN A., LOWE P. (1994) Local politics and rural restructuring: the case of the contested countryside. In: SYMES D., JANSEN A.J. (eds) *Agricultural Restructuring and Rural Change in Europe* (Wageningen Agricultural University, Wageningen), pp. 247-259.

FLYNN A., MARSDEN T.K. (1992) Food regulation in a period of agricultural retreat: the British experience *Geoforum* 23, pp. 85–93.

FLYNN A., MARSDEN T.K. (1995) Rural change, regulation and sustainability *Environment and Planning A* 27, pp. 1180–1193.

FLYNN A., MARSDEN T.K., WARD N. (1994) Retailing, the food system and the regulatory state. In: LOWE P., MARSDEN T.K., WHATMORE S. (eds) *Regulating Agriculture* (David Fulton Publishers, London)

FRANKLIN A (2000) Institutional suitability: establishing an effective structural framework for integrated rural development. Paper presented to the European Society for Rural Sociology Summer School, Finland.

FRIEDMANN H, MCMICHAEL P. (1989) Agriculture and the state system: the rise and decline of national agricultures, 1870 to present *Sociologia Ruralis* 28, pp. 93–118.

GARROD N. (2000) Environmental contingencies and sustainable modes of corporate governance *Journal of Accounting and Public Policy* 19, pp. 237–261.

GIBBS D. (2001) Globalisation, the bioscience industry and local environmental responses *Global Environmental Change* 10, pp. 245–257.

GLIESSMAN S.R. (1990) *Agro-ecology. Researching the Ecological Basis for Sustainable Agriculture* (Springer Verlag, New York).

GOMES DA SILVA A. (1993) Irrigação x força de trabalho: escassez ou resistência. In: XIMENES T. (ed.) *Novos Paradigmas e Realidade Brasileira* (ANPOCS/UFPA AEA, Belém), pp. 338–355.

GOODMAN D. (1999) Agro-food studies in the 'age of ecology': nature, corporeality, bio-politics *Sociologia Ruralis* 39, 17–38.

GOODMAN D. (2001) Ontology matters : the relational materiality of nature and agro-food studies *Sociologia Ruralis* 41, 182–201.

GOODMAN D., WATTS M. (1994) Reconfiguring the rural or fording the divide? Capitalist restructuring and the global agro-food system *Journal of Peasant Studies* 22, pp. 1–49.

GRABHER G. (ed.) (1993) *The Embedded Firm. On the Socioeconomics of Industrial Networks* (Routledge, London).

GRAHAM V. (2000) *The EU and GM Foods. Current Regulations and Future Trends* (Chando Publishing, Oxford).

GRAY J. (2000) The Common Agricultural Policy and the re-invention of the rural in the European Community *Sociologia Ruralis* 40, pp. 30–52.

GRAY R., WALTERS D., BEBBINGTON J., THOMPSON I. (1995) The greening of enterprise: an exploration of the (NON) role of environmental accounting and accountants in organizational change *Critical Perspectives on Accounting* 6, pp. 211–239.

GUY C., KITCHEN L., MARSDEN T., MURDOCH J. (2000) Pilot study of local retailer-supplier linkages in two rural areas. Unpublished report for Tesco Stores Ltd.

GUZMAN E., WOODGATE G. (1999) Alternative food and agriculture networks: an agro-ecological perspective on the responses to economic globalisation and the 'new agrarian question'. Paper presented at EU COST Workshop.

HALKIER B. (1999), Consequences of the politicization of consumption: the example of environmentally friendly consumption practices *Journal of Environmental Policy and Planning* 1, pp. 25–41.

HALL P., THOMAS R., GRACEY H., DREWELT R. (1973) *The Containment of Urban England* (Allen & Unwin, London).

HANCHER L., MORAN M. (1989) Organising regulatory space. In: HANCHER L., MORAN M. (eds) *Capitalism, Culture and Economic Regulation* (Clarendon Press, Oxford), pp. 271-299.

HARVEY D. (1996) The role of markets in the rural economy. In: ALLANSON P., WHITBY M. (eds) *The Rural Economy and the British Countryside* (Earthscan Publications, London) pp. 19–40.

HECHT S. (1987) The evolution of agroecological thought. In: ALTIERI MA (eds) *Agroecology. The Scientific Basis of Alternative Agriculture* (IT Publications, London), pp. 1-20.

HILL S., ROBERTS A. (1998) *Welsh Input–Output Tables for 1995* (University of Wales Press, Cardiff).

HIRSCH F. (1976) *Social Limits to Growth* (Harvard University Press, Cambridge, MA).

HOC GRAND COMM (HOUSE OF COMMONS WELSH GRAND COMMISSION) (1962) *Minutes of Proceedings, 30 May 1962, Agriculture in Wales* (HMSO, London).

HOC GRAND COMM (HOUSE OF COMMONS WELSH GRAND COMMISSION) (1968) *Minutes of Proceedings, 3 April 1968, Agriculture in Wales* (HMSO, London).

HOWELL D. (1977) *Land and People in Nineteenth Century Wales* (Routledge & Kegan Paul, London).

IACOPONI L., BRUNORI G., ROVAI M. (1995) Endogenous development and the agro-industrial district. In: PLOEG J.D. VAN DER, DIJK G. VAN (eds) *Beyond*

Modernization. The Impact of Endogenous Rural Development (Van Gorcum, Assen), pp. 28-69.

IRWIN A. (2001) *Sociology and the Environment* (Polity Press, Oxford).

JACOBS H.M. (1999) Who owns America? *University of Wisconsin–Madison Land Tenure Center Newsletter* 77, pp. 6–7.

JORDAN A., RICHARDSON J. (1987) *Government and Pressure Groups in Britain* (Clarendon Press, Oxford).

KALTOFT P. (1999) Values about nature in organic farming practice and knowledge *Sociologia Ruralis* 39, pp. 39–53.

KEIL R. (2000) Discussants comments at the 'Natural spaces? Governing spaces of nature in the new economy' session. American Geographers Annual Meeting, Pittsburgh, 4–8 April 2000.

KINSELLA J., MANNION J., BOGUE P. (2000) *Creating a New Future for Dairy Farm Households: A Key Element of Rural Renewal. County Clare Partnership Dairy Action Research Programme 1997–9* (University College Dublin, Dublin).

KITCHEN L. (2000) Environmental policy and the differentiation of rural space: an actor-network perspective *Journal of Environmental Policy and Planning* 2, pp. 135–147.

KNEALE J., LOWE P., MARSDEN T.K. (1991) *The Conversion of Agricultural Buildings.* ESRC Countryside Change Initiative Working Paper 29 (University of Newcastle, Newcastle).

LAMPKIN N., MEASURES M. (eds) (2000) *Organic Farm Management Handbook* (3rd edn) (Welsh Institute of Rural Studies, University of Wales, Aberystwyth).

LAPPE M., BAILEY B. (1999) *Against the Grain. The Genetic Transformation of Global Agriculture* (Earthscan Publications, London).

LATOUR B. (1987) *Science in Action. How to Follow Scientists and Engineers through Society* (Open University Press, Milton Keynes).

LAW J. (1992) Notes on the theory of the actor-network: ordering, strategy and heterogeneity *Systems Practice* 5, pp. 379–393.

LIGHT A. (ed.) (1998) *Social Ecology after Bookchin* (Guildford Press, New York).

LIMA P. (1994) Economia do Nordeste: tendências recentes da áreas dinâmicas. Textos para discussão (PIMES–UFPE, Recife).

LIPSCHUTZ R.D. (1999) From local knowledge and practice to global environmental governance. In: HEWSON, M., SINCLAIR T.J. (eds) *Approaches to Global Governance Theory* (SUNY Press, Albany, NY), pp. 259-283.

LLAMBI L. (1993) Globalización y nueva ruralidad en America Latina: una agenda teorica de investigación *Revista da ALASRU* 2, 29–39.

LONG N. (1996). Globalization and localization: new challenges to rural research. In: MOORE H.L. (ed.) *The Future of Anthropological Knowledge. The Uses of Knowledges: Global and Local Relations* (Routledge, London), pp. 37-59.

LONG N. (2001) *The Power of Music. Issues of Agency and Social Practice* (Wageningen University, Wageningen).

LONG N., LONG A. (eds) (1992) *Battlefields of Knowledge: the Interlocking of Theory and Practice in Social Research and Development* (Routledge, London).

LONG N., PLOEG J.D. VAN DER (1995) Reflections on agency, ordering the future, and planning. In: FRERKS G., OUDEN J.H.B. DEN (eds) *In Search of the Middle Ground. Essays on the Sociology of Planned Development* (Wageningen Agricultural University, Wageningen), pp. 64-78.

LOWE P., MARSDEN T.K., WHATMORE S. (eds) (1990) *Technological Change and the Rural Environment* (David Fulton Publishers, London).

LOWE P., MURDOCH J., MARSDEN T.K., MUNTON R., FLYNN A. (1993) Regulating the new rural spaces: the uneven development of land *Journal of Rural Studies* 9, pp. 205–223.

LOWE P., MURDOCH J., WARD N. (1995) Networks in rural development: beyond exogenous and endogenous models. In: PLOEG J.D. VAN DER, DIJK G. VAN (eds) *Beyond Modernization. The Impact of Endogenous Rural Development* (Van Gorcum, Assen) pp. 87–105.

LOWE P., CLARK J., SEYMOUR S., WARD N. (1997) *Moralising the Environment. Countryside Change, Farming and Pollution* (UCL Press, London).

LURY C. (1996) *Consuming Culture* (Polity Press, Cambridge).

MACAULEY D. (1999) Evolution and revolution: the ecological anarchism of Kropotkin and Bookchin. In: LIGHT A. (ed.) *Social Ecology after Bookchin* (Guildford Press, New York).

MACKERRON D.K.L., DUNCAN J.M., HILLMAN J.R., MACKAY G.R., ROBINSON D.J., TRUDGILL D.L., WHEATLEY R.J. (2000) *Organic Farming Science and Belief.* (Scottish Crop Research Institute, http://www.scri.sari.ac.uk)

MAFF (MINISTRY OF AGRICULTURE, FISHERIES AND FOOD) (1998) *Review of the Organic Aid Scheme*
(MAFF, http://www.maff.gov.uk/environ/envsch/review.doc).

MAFF (MINISTRY OF AGRICULTURE, FISHERIES AND FOOD) (1999) *The Future of Agriculture in England.*

MAI L.W., NESS M. (1998) Perceived benefits of mail-order speciality food *British Food Journal* 100(1), pp.10–17.

MANN S., DICKINSON J. (1980) State and agriculture in two eras of American capitalism. In: BUTTEL F., NEWBY H. (eds) *The Rural Sociology of Advanced Societies. Critical Perspectives* (Allanheld, Osmun & Co., Montclair, NJ) pp. 283–325.

MARSDEN T.K. (1991) Theoretical issues in the continuity of petty commodity production. In: WHATMORE S., LOWE P., MARSDEN T. (eds) *Rural Enterprise. Shifting Perspectives on Small-scale Production* (David Fulton Publishers, London).

MARSDEN T.K. (1992) Exploring a rural sociology for the Fordist transition. Incorporating social relations into economic restructuring *Sociologia Ruralis* 32, pp. 209–231.

MARSDEN T.K. (1995) Beyond agriculture: regulating the new rural spaces *Journal of Rural Studies* 11, pp. 285–297.

MARSDEN T.K. (1997) Institutions and structural organisation in Rural Wales. Paper presented to the Institute of Welsh Affairs annual conference, Unversity of Wales Aberystwyth, July.

MARSDEN T.K. (1999) Rural futures: the consumption countryside and its regulations *Sociologia Ruralis* 39, pp. 501–521.

MARSDEN T. (forthcoming) The Rural Sociology of Food: Understanding the Supply and Consumption of Food and Drink.

MARSDEN T.K., ARCE A. (1995) Constructing quality: emerging food networks in the rural transition *Environment and Planning A* 27, pp. 1261–1279.

MARSDEN T., WRIGLEY N. (1995) Regulation, retailing, and consumption *Environment and Planning A* 27, pp. 1899–1912.

MARSDEN T., WRIGLEY N. (1996) Retailing, the food system and the regulatory state. In: WRIGLEY N., LOWE M.S. (eds) *Retailing, Consumption and Capital. Towards the New Retail Geography* (Longman, Harlow) pp. 33–47.

MARSDEN T.K., MUNTON R., WHATMORE S., LITTLE J. (1986) Towards a political economy of capitalist agriculture: a British perspective *International Journal of Urban and Regional Research* 10, pp. 498–521.

MARSDEN T.K., MURDOCH J., LOWE P., MUNTON R., FLYNN A. (1993) *Constructing the Countryside* (UCL Press, London).

MARSDEN T.K., CAVALCANTI J.S.B., FERREIRA IRMAO J. (1996a) Globalisation, regionalisation and quality: the socio-economic reconstituion of food in the San Francisco Valley, Brazil *International Journal of Sociology and Food* 5, pp. 85–114.

MARSDEN T.K., MUNTON R., WARD N., WHATMORE S. (1996b), Agricultural geography and the political economy approach: a review *Economic Geography* 72, pp. 361–375.

MARSDEN T.K., FLYNN A., HARRISON M. (1998) Creating competitve space: exploring the social and political maintenance of retail power *Environment and Planning A* 30, pp. 481–498.

MARSDEN T.K., MURDOCH J., MORGAN K. (1999) Food supply chains and regional development *International Planning Studies* 4, pp. 295-302.

MARSDEN T.K., BANKS J., BRISTOW G. (2000a) Food supply chain approaches: exploring their role in rural development *Sociologia Ruralis* 40, pp. 424–438.

MARSDEN T., FLYNN A., HARRISON M. (2000b) *Consuming Interests. The Social Provision of Foods* (UCL Press, London).

MARSDEN T., BANKS J., RENTING H., PLOEG J.D. VAN DER (2001a) The road towards sustainable rural development: issues of theory, policy and research practice *Journal of Environmental Policy & Planning* 3, pp. 75–85.

MARSDEN T.K., MILBOURNE P., KITCHEN L., BISHOP K. (2001b) Communities in Nature: the construction and understanding of forest natures in South Wales Valleys Communities. Paper at the XIX European Congress for Rural Sociology, Dijon, September, 2001.

MARSH J. (1997) *A Review of the Role of Agriculture, Horticulture and Forestry in the UK Economy* (Office of Science and Technology, London).

MARTIN R., MARSDEN T. (1999) Food for urban spaces: the development of urban food production in England and Wales *International Planning Studies* 4, pp. 389–413.

MATHER A.S. (2001) Forests of consumption: postproductivism, postmaterialism, and the postindustrial forest *Environment and Planning C: Government and Pollicy* 19, 249–268.

MCDONAGH P., COMMINS P. (1999) Food chains, small scale enterprises and rural development: illustrations from Ireland *International Planning Studies* 4, pp. 349–373.

MCMICHAEL P. (2000) The power of food *Agriculture and Human Values* 17, pp. 21–33.

MENON A., MENON A. (1997) Entreprenuerial marketing strategy: the emergence of corporate environmentalism as market strategy *Journal of Marketing* 61, pp. 51–67.

MICHELSEN J. (1995) Organic products in conventional distribution systems: experiences from Denmark. Mimeo, South Jutland Research Centre, Esbjerg.

MICHELSEN J., HAMM U., WYNEN E., ROTH E. (1999) *The European Market for Organic Products. Growth and Development* (Universität Hohenheim, Stuttgart).

MIDMORE P., SHERWOOD A.-M., ROUGHLEY G. (2001) Policy reform and the sustainability of farming in the uplands of the United Kingdom: conflicts between environment and social support *Journal of Environmental Policy and Planning* 3, pp. 43–63.

MILLER D. (1995) *Worlds Apart. Modernity through the Prism of the Local* (Routledge, London).

MINISTRY OF WORKS AND PLANNING (1942) *Report of the Committee on Land Utilisation in Rural Areas* (HMSO, London).

MOL A.P.J. (1999) Ecological modernization and the environmental transition of Europe: between national variations and common denominators *Journal of Environmental Policy and Planning* 1, pp. 167–181.

MOONEY P. (1985), *My Own Boss? Class, Rationality, and the Family Farm* (Westview Press, Boulder, CO).

MORGAN K., MUNGHAM G. (2000) *Redesigning Democracy. The Making of the Welsh Assembly* (Seren, Bridgend).

MORMONT M. (1990) Who is rural? Or, how to be rural: towards a sociology of the rural. In: MARSDEN T. K., LOWE P., WHATMORE S. (eds) *Rural Restructuring. Global Processes and Local Responses* (London, David Fulton Publishers).

MUNTON R. (1995) Regulating rural change: property rights, economy and environment: a case study from Cumbria, UK *Journal of Rural Studies* 11, pp. 267–284.

MURDOCH J. (1988) The State and Agriculture in Rural Wales. Unpublished PhD thesis (The University of Wales, Aberystwyth).

MURDOCH J. (1992) Representing the region: Welsh farmers and the British state. In MARSDEN T., LOWE P., WHATMORE S. (eds) *Labour and Locality. Uneven Development and the Rural Labour Process.* (David Fulton Publishers, London), pp. 160–180.

MURDOCH J. (2001) Ecologising sociology: actor network theory, co-construction and the problem of human exceptionalism *Sociology* 35, pp. 111–133.

MURDOCH J., ABRAM S. (1997) Defining the limits of community governance *Journal of Rural Studies* 14, pp. 41–50.

MURDOCH J., MARSDEN T.K. (1994) *Reconstituting Rurality. Class, Community and Power in the Development Process* (UCL Press, London).

MURDOCH J., MARSDEN T.K. (1995) The spatialization of politics: local and national actor-spaces in environmental conflict *Transactions of the Institute of British Geographers* 20, pp. 368–380.

MURDOCH J., MIELE M. (1999) 'Back to nature': changing 'worlds of production' in the food sector *Sociologia Ruralis* 39, pp. 465–483.

MURDOCH J., MORGAN K. (1996) *Exploring the 'Third Way'. Networks in European Rural Development* (OECD, Paris).

MURDOCH J., MARSDEN T.K., BANKS J. (2000) Quality, nature, and embeddedness: some theoretical considerations in the context of the food sector *Economic Geography* 76, pp. 107–125.

MURMIS M. (1993) Algunos temas para la discusión en la sociología rural latinoamericana: rees-tructuración, desestructuración y problemas de excluidos e incluidos *Revista da ALASRU* 2, pp. 5–28.

NAW (NATIONAL ASSEMBLY FOR WALES) (1999a) *Rural Development Plan for Wales 2000–2006* (National Assembly for Wales, Cardiff).

NAW (NATIONAL ASSEMBLY FOR WALES) (1999b) *Rural Wales. A Statement by the Rural Partnership* (National Assembly for Wales, Cardiff).

NAW (NATIONAL ASSEMBLY FOR WALES) (2001a) *Diversifying the Rural Economy* (National Assembly for Wales, Cardiff).

NAW (NATIONAL ASSEMBLY FOR WALES) (2001b) *Farming for the Future* (National Assembly for Wales, Cardiff).

NEWBY H., BELL C., ROSE D., SAUNDERS P. (1978) *Property, Paternalism and Power. Class and Control in Rural England* (Hutchinson, London).

NFU (NATIONAL FARMERS UNION) (2000) *Farmers Markets Business Survey* (NFU Press Office, London).

NIX J. (2000) *Farm Management Handbook* (31st edn) (Imperial College at Wye, Wye).

NORGAARD R. (1994) *Development Betrayed* (Routledge, London).

NYGARD B., STORSTAD O. (1998) De-globalization of food markets? Consumer perceptions of safe food: the case of Norway *Sociologia Ruralis* 38, pp. 35–53.

OECD (ORGANISATION FOR ECONOMIC CO-OPERATION AND DEVELOPMENT) (1989) *Agricultural and Environmental Policies. Opportunities for Integration* (OECD, Paris).

PANZAR J., WILLIG R. (1981) Economies of scope *American Economic Review* 71, pp. 268–272.

PARROTT N., MARSDEN T.K. (2001) *'The Real Green Revolution'. Organic and Agro-ecological Farming in the South* (Greenpeace Trust, London).

PEATTIE K. (1999) *Environmental Marketing Management: meeting the green challenge* (Pitman Publishing, London).

PICCHI A. (1994) The relations between central and local powers as context for endogenous development. In: PLOEG J.D. VAN DER, LONG A. (eds) *Born from Within. Practice and Perspectives of Endogenous Development* (Van Gorcum, Assen).

PLOEG J.D. VAN DER (1997), On rurality, rural development and rural sociology. In: HAAN H. DE, LONG N. (eds) *Images and Realities of Rural Life. Wageningen Perspectives on Rural Transformation* (Van Gorcum, Assen) pp. 39–73.

PLOEG J.D. VAN DER, DIJK G. VAN (eds) (1995) *Beyond Modernization. The Impact of Endogenous Rural Development* (Van Gorcum, Assen).

PLOEG J.D. VAN DER, SACCOMANDI V. (1995) On the impact of endogenous development in agriculture. In: PLOEG J. D. VAN DER, DIJK G. VAN (eds) *Beyond Modernization. The Impact of Endogenous Rural Development* (Van Gorcum, Assen) pp. 10–27.

PLOEG J.D. VAN DER, RENTING H., BRUNORI G., KNICKEL K., MANNION J., MARSDEN T., ROEST K. DE, SEVILLA GUZMÁN E., VENTURA F. (2000) Rural development: from practices and policies toward theory *Sociologia Ruralis* 40, pp. 391–408.

POLANYI K. (1944) *The Great Transformation. The Political and Economic Origins of Our Time* (Beacon Press, Boston, MA).

POLONSKY M.J. (1995) A stakeholder theory approach to designing environmental marketing strategy *Journal of Business and Industrial Marketing* 10(3), pp. 29–46.

RABOBANK NEDERLAND (1993) *The World Fresh Fruit Market* (Rabobank Nederland Agribusiness Research, Utrecht).

RABOBANK NEDERLAND (1994) *The Retail Food Market. Structure, Trends and Strategies* (Rabobank Nederland Agribusiness Research, Utrecht).

RANDALL P. (1972) Wales in the structure of central government *Public Administration* 50, pp. 353–373.

REDCLIFT M., WOODGATE G. (1998) (eds) *The International Handbook of Environmental Sociology* (Edward Elgar Publishing, Cheltenham).

SACCOMANDI V. (1998) *Agricultural Market Economics. A Neo-institutional Analysis of the Exchange, Circulation and Distribution of Agricultural Products* (Van Gorcum, Assen).

SAINSBURY'S SUPERMARKETS (1997) *Sainsbury's and the Environment. Organic Food* (Sainsbury's Supermarkets, London).

SALAIS R., STORPER M. (1992) The four worlds of contemporary industry. *Cambridge Journal of Economics* 16, pp. 169–193.

SAUNDERS P., HARRIS C. (1990) Privatisation and the consumer *Sociology* 24, pp. 57–75.

SDRN (SUSTAINABLE DEVELOPMENT RESEARCH NETWORK) (2002) *A New Agenda for UK Sustainable Development Research* (Policy Studies Institute, London).

SEVILLA GUZMÁN E., WOODGATE G.R. (1997) Sustainable rural development: from industrial agriculture to agroecology. In: REDCLIFT M., WOODGATE G.R. (eds) *The International Handbook of Environmental Sociology* (Edward Elgar Publishers, Cheltenham).

SEVILLA GUZMÁN E., WOODGATE G.R. (1999) From farming systems research to agroecology. In: *Technical and Social Systems Approaches for Sustainable Rural Development*. European Commission. Report 45/98 (European Commission, Brussels).

SHUCKSMITH M. (2000) Social Exclusion in Rural Areas (Joseph Rowntree Research Trust, York).

SMITH R.J. (1992) On the margins: uneven development and rural restructuring in the Highlands of Scotland. In MARSDEN T., LOWE P., WHATMORE S. (eds) *Labour and Locality. Uneven Development and the Rural Labour Process* (David Fulton Publishers, London) pp. 136–159.

SOIL ASSOCIATION (2000) *Multiple Retailers and Organic Foods*. Briefing Paper (Soil Association, http://www.soilassociation.org).

SOIL ASSOCIATION (2001) *Organic facts and figures - February 2001* (Soil Association, http://www.soilassociation.org).

STORPER M. (1997) *The Regional World* (Guildford Press, London).

TARLING R., RHODES J., NORTH J., BROOM G. (1993) *The Economy and Rural England* (Rural Development Commission, London).

THEVENOT L. (1998) Innovating in 'qualified' markets: quality, norms and conventions. Paper presented at Systems and Trajectories of Innovation Conference, Institute of International Studies, University of California, Berkeley, April 1998.

TOLEDO V. (1989) The ecological rationality of peasant production. In: ALTIERI M., HECHT S. (eds) *Agroecology and Small Farm Development* (CRC Press, Boca Raton, FL), pp. 53–60.

TORDJMAN A. (1995) *Trends in Europe. Consumer Attitudes and the Supermarket 1995* (Food Marketing Institute, Washington, DC).

TOVEY H. (1997) Food, environmentalism and rural sociology: on the organic farming movement in Ireland *Sociologia Ruralis* 37, pp. 21–37.

UFPE. PIME (1991) *Impactos econômicos da irrigação sobre o pólo Petrolina Juazeiro* (Editora Universitaria da UFPE, Recife).

URRY J. (1995) *Consuming Places* (Routledge, London).

WARD N. (1990) A preliminary analysis of the UK food chain *Food Policy* 15, pp. 439–441.

WARD N., LOWE P., SEYMOUR S., CLARK J. (1995) Rural restructuring and the regulation of farm pollution *Environment and Planning A* 27, pp. 1193–1213.

WATTS M. (1996) Development III. The global agro-food system and late twentieth century development (or Kautsky redux) *Progress in Human Geography* 20, pp. 230–245.

WEBER M. (1949) *The Methodology of the Social Sciences*, SHILS E.A., FINCH H.A. (eds) (Free Press, New York).

WELSH OFFICE AGRICULTURE DEPARTMENT (1998) Evidence to the Welsh Affairs Committee Inquiry into the problems facing the livestock industry. In: HOUSE OF COMMONS WELSH AFFAIRS COMMITTEE *2nd report [session 1997-98] : the present crisis in the Welsh livestock industry : report, together with the proceedings of the Committee, minutes of evidence and appendices* (Stationery Office, London).

WHATMORE S., MUNTON R., MARSDEN T. (1990) The rural restructuring process: emerging diversions of property rights *Regional Studies* 24, pp. 235–245.

WILLIAMS L.J. (1983) The economic structure of Wales since 1859. In WILLIAMS G. (ed.) *Crisis of Economy and Ideology. Essays on Welsh Society, 1840–1980* (BSA, Bangor).

WILLIAMS L.J. (1985) *Welsh Historical Statistics* (2 vols) (Welsh Office, Cardiff).

About the author

Terry Marsden is Professor of Environmental Policy and Planning and Head of Department in the Department of City and Regional Planning, Cardiff University, Wales, UK. He is also Co-Director of the ESRC Research Centre 'Business Relationships, Accountability, Sustainability and Society' (BRASS) at Cardiff.